WILDLIFE WEEKENDS

IN SOUTHERN BRITISH COLUMBIA

WILDLIFE WEEKENDS
IN SOUTHERN BRITISH COLUMBIA

DAY AND MULTI-DAY TRIPS FROM VANCOUVER FOR WILDLIFE VIEWING

ROY JANTZEN

With Contributions by
Patricia Thomson, Dr. Robert Butler, Dr. Richard Kool, Anne Murray, and Johanna Wagstaffe

RMB
Think outside.

First Edition

For information on purchasing bulk quantities of this book, or to obtain media excerpts or invite the author to speak at an event, please visit rmbooks.com and select the "Contact" tab.

RMB | Rocky Mountain Books Ltd.
rmbooks.com
@rmbooks
facebook.com/rmbooks

Cataloguing data available from Library and Archives Canada
ISBN 9781771604604 (paperback)
ISBN 9781771604611 (electronic)

Cover and interior layout by Lara Minja, Lime Design
Maps and charts designed by Shannon ten Brink

Printed and bound in China

We would like to take this opportunity to acknowledge the Traditional Territories upon which we live and work. In Calgary, Alberta, we acknowledge the Niitsítapi (Blackfoot) and the people of the Treaty 7 region in Southern Alberta, which includes the Siksika, the Piikuni, the Kainai, the Tsuut'ina, and the Stoney Nakoda First Nations, including Chiniki, Bearpaw, and Wesley First Nations. The City of Calgary is also home to Métis Nation of Alberta, Region III. In Victoria, British Columbia, we acknowledge the Traditional Territories of the Lkwungen (Esquimalt and Songhees), Malahat, Pacheedaht, Scia'new, T'Sou-ke, and W̱SÁNEĆ (Pauquachin, Tsartlip, Tsawout, Tseycum) Peoples.

We acknowledge the financial support of the Government of Canada through the Canada Book Fund and the Canada Council for the Arts, and of the province of British Columbia through the British Columbia Arts Council and the Book Publishing Tax Credit.

Disclaimer

The actions described in this book may be considered inherently dangerous activities. Individuals undertake these activities at their own risk. The information put forth in this guide has been collected from a variety of sources and is not guaranteed to be completely accurate or reliable. Many conditions and some information may change owing to weather and numerous other factors beyond the control of the authors and publishers. Individuals or groups must determine the risks, use their own judgment, and take full responsibility for their actions. Do not depend on any information found in this book for your own personal safety. Your safety depends on your own good judgment based on your skills, education, and experience.

It is up to the users of this guidebook to acquire the necessary skills for safe experiences and to exercise caution in potentially hazardous areas. The authors and publishers of this guide accept no responsibility for your actions or the results that occur from another's actions, choices, or judgments. If you have any doubt as to your safety or your ability to attempt anything described in this guidebook, do not attempt it.

To my wife Heather,

for your support and guidance throughout the research, writing, and publishing of this book. You were my co-interviewer, co-pilot, navigator, my note taker, my editor, re-editor, re-re-editor; you were the one who deconstructed, reconstructed, cut and pasted, fact-checked, and added complementary ideas, while making those of mine more clearly understood. Thank you for helping get this book to completion; I could not have done it without you!

Love, Roy

I acknowledge with respect the unceded Traditional Territories upon which I live, work, and write. From my home in Port Moody, BC, I acknowledge specifically the Kwikwetlem, Tsleil-Waututh (səl̓ilwətaʔɬ təməxʷ), Stó:lō (S'ólh Téméxw), Coast Salish, Squamish (Skwxwú7mesh-ulh Temíx̱w), Qayqayt, and Musqueam (xʷməθkʷəy̓əm) Peoples.

In researching the traditional First Nation territories for all viewing locations outlined in this book, I referred to the site native-land.ca to guide me.

CONTENTS

COYOTES ARE VERY AWARE OF THEIR SURROUNDINGS, SO IF YOU HAPPEN TO SPOT ONE, IT WILL LIKELY MAKE EYE CONTACT AS IT ASSESSES YOUR INTENTIONS.
PHOTO ROB ALEXANDER.

ACKNOWLEDGEMENTS

A healthy ecosystem exists only when all of the individuals work together to form another kind of "organism," one that is greater than the sum of its parts. To me, this book project felt like an ecosystem of sorts, and the individuals who contributed to it represented a wide variety of scientific disciplines, from lichenology to cetology. I consider myself a wildlife generalist, focused on interpretation and education, with any perceived expertise attained by riding on the coattails of generous specialists. For this project, I was the beneficiary of the wisdom of a multitude of wildlife researchers, comprising over 50 interviewees, each freely sharing their time and immense knowledge with me, making this guidebook possible.

The collaborative process also included those who have generously shared their incredible wildlife photographs to accompany the text. Those who photograph wild creatures notice the most miniscule details and have so much knowledge to share about the animals, beyond the images themselves. I am in awe of the stories they tell.

I am indebted to all of them.

Special thanks to...

My wife Heather, who supported me throughout this process.

Frank Ritcey, for multiple interviews and discussions, for two separate hikes and an invitation for more, for content editing and proofreading (and for teaching me about the Oxford comma), and for working with me to ensure the chapters on Black Bears, Bighorn Sheep, and Moose were accurate. Frank, whom I did not know before this project, has become an integral part of it on my behalf.

The writing contributors who helped me directly include the voices that have influenced me over time: Patricia Thomson, Dr. Rob Butler, Dr. Richard Kool, Johanna Wagstaffe, and Anne Murray.

Thank you to the experts and interviewees without whose input this book would not have been possible:

Great Blue Herons: Camille Coray (interview, content editing, images); Interviews with: Ross Vennesland, Fernando Selles, Abbey Riddolls, and Eric Gross

Hummingbirds: Sue Elwell (visit, interview, images, and content editing); Dr. Alison Moran (guidance, interview, images, and content editing)

Black Bears: Trevor Goward (interview and inspirational ideas); Bob Hansen (interview and networking); Mike Mueller (interview)

Meadowlark Nature Festival: Dr. Janet Willson (interview and content editing)

Vancouver Island Marmots: Dr. Malcolm McAdie (multiple interviews and content editing); Adam Taylor (interview, content editing, and images); Doug Routledge (interview and inspirational ideas)

Insects: Denis Knopp (interviews); Crispin Guppy (interview and guidance); Winnie Hwo (interview); Syd Cannings (interview, content editing, and helping me with a key style choice); Rob Cannings (content editing)

Snakes: Dr. Christine Bishop and Dana Eye (interview); Barb Sabyan (networking and images); Dr. Karl Larsen (interview); Chloe Howarth (interview, content editing, and images); Stephanie Winton (interview)

Small Mammals: Kelly Pearce (interview, content editing, and images); Jo Hughes (interview and content editing)

Bats: Darlene Hartford (hosting, interview, content editing, networking and the BEEPS Board of Directors, Heidi, Rachel, Dagmar, Randi, Barb, and Nicola); Tanya Luszcz (interview); Dr. Cori Lausen (guidance); John Saremba (interview and images); Ella Braden and Tina Deenik (interviews)

Coyotes: Dannie Piezas (interview, content editing, networking, and images); Dan Straker (interview, content editing, and networking); Kristen Walker (interview); Jay White (interview)

Intertidal: Phil Lavoie (interview and content editing)

Orcas: Lucy Quayle (interview, content editing, networking, and images); Bob Bruce (interview and networking); Dr. Martin Wale (interview and content editing); Donna Sandstrom (interview, graphics, and content editing); Susie Washington-Smyth (interview)

Turkey Vultures: Ann Nightingale (interview, networking, and content editing)

Bighorn Sheep: Edyta Marcisz (interview, networking, and content editing)

Bald Eagles: Dr. Judith Knapp and Wolfgang Pauli (interview, content editing, and images); David Hancock (interview and content editing)

Grey Whales: Carla Moss (interview, networking); Tom Balfour (interview); Todd Windle (interview); Moses Martin (interview and guidance); Jim Darling (interview and content editing); Laura Loucks (interview)

Herring: Scott Wallace (interview and content editing); Cath Gray (content editing)

Winter Birds: Anne Murray (content editing)

Thank you to all those who helped guide me, inform me, and connect me with others. These include Cori Lausen, Jared Hobbs, Robyn Worcester, Kathleen Stormont, Dr. Alan Burger, Andrea Lawrence, Christian Engelstoft, and Jesse Ritcey.

Thank you to the experts and interviewees without whose input this book would not have been possible:

It is the images that make a book visually appealing, so I am deeply grateful to all of the photographers who contributed photos. The following photographers provided *multiple* images for *multiple* chapters:

Rob Alexander – Rob, the quiet photographer with an amazing eye for detail, has taught me how to look deeply at species and focus on their astounding, sometimes barely visible, adaptations.

Terry Carr – Terry was central to pulling together images for this project – his own and from others affiliated with the Delta Naturalists Society. As someone I didn't know prior to this project, he exemplifies how gracious and generous people can be.

Sabine Decamp – I was connected to Sabine through a friend common to both of us. When she saw what I was looking for, she shared multiple photo albums with me, with images of the coastal species that appear in this book. She is an incredibly talented photographer and outdoorswoman.

Christine Gaio – Christine appeared to have a photo for almost every idea I came up with! She is an incredibly talented photographer and I am grateful she supports my work so enthusiastically.

Eric Gross – I interviewed Eric about Great Blue Herons, and, when a common friend heard I was looking for images for the book, I was reconnected with him. He, too, was someone who provided an array of photos connected to multiple chapters in this book.

Michael Schmidt – I met Michael late in this process, after the Stanley Park Ecology Society sent me a photo of a Coyote with a Squirrel in its mouth. I reached out to him to ask for approval to use the image, and he suggested I look at michaelschmidtphotography.com to see if there were other images that would suit the text. He offered to contribute as many as I needed. The composition, lighting, and emotion were stunning in so many of his photos, it was hard to choose.

Others who answered a callout for specific species images to accompany the text (not mentioned above) include Jon Moran, Marion Shikaze, D. Hildebrand, Walter Sussel, Mike Lester, Jordan Cormack, Pascale Charland, Jade Spruyt, Milva DeSiena, Mark White, Phil Lavoie, Jordan Hawkswell, Jeanne Hyde, Terry Berezan, Linda at Brakelightfoto.com, Tym Cyr, Don Travers, Shelley Frick, Deborah Simpson, Chris McVittie, and Cameron Eckert.

Several of my colleagues at Capilano University supported me in the process of getting this project off the ground. They include Cheryl Schreader, Tom Flower, Bob Muckle, Stephanie Wells, Catherine Boniface, Caroline Dépatie, Brad Martin, and Greig Gjerdalen.

Thank you to Shannon Tenbrink, a superb graphic designer, who at a late stage in the process jumped in to ensure completion of many maps and graphics for this book.

For about the last decade, as the concept of this book was solidifying in my mind, there were those whom I reached out to for their ideas and feedback, and it was through these conversations that many of the ideas were developed. These people include Terry and Evelyne Berezan, David Henry, Catherine Evans, Scott and Carol Fisher, and John and Sue Glynn-Morris.

Thank you to Patricia Thomson who not only wrote the foreword to this book, but also proofread the complete book from a naturalist's perspective to help ensure the clarity of the writing and the ideas they contain.

THE 20 WILDLIFE VIEWING LOCATIONS IN THIS BOOK ARE WITHIN A SIX-HOUR DRIVE OF VANCOUVER AND ARE DESIGNED TO BE VISITED IN AS LITTLE AS ONE, OR UP TO FOUR DAYS.

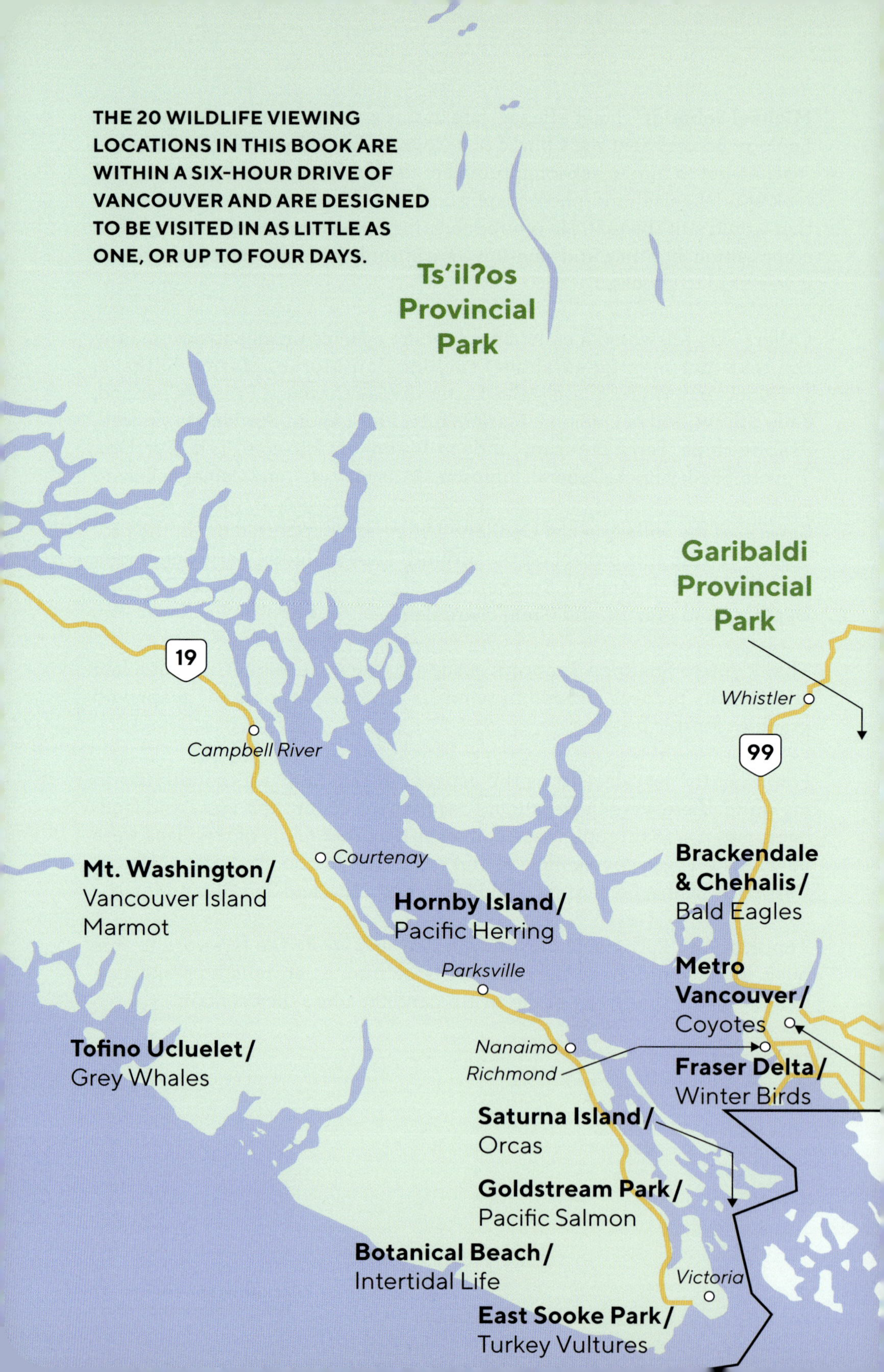

Wells Gray Park / Black Bears & Moose
100 Mile House
Clearwater
97
5
99
Lac du Bois / Bighorn Sheep
Sun Peaks
1
Salmon Arm
Kamloops
97
Vernon
Peachland / Bats
5
5A
97C
Kelowna
5
Penticton
Princeton / Hummingbirds
Skagit Valley / Butterflies & Dragonflies
South Okanagan / Meadowlark Festival
7
Chilliwack / Great Blue Heron
3
Manning Park / Small Mammals
Cathedral Park / Mountain Goats
Osoyoos / Western Rattlesnake
Abbotsford
Vancouver

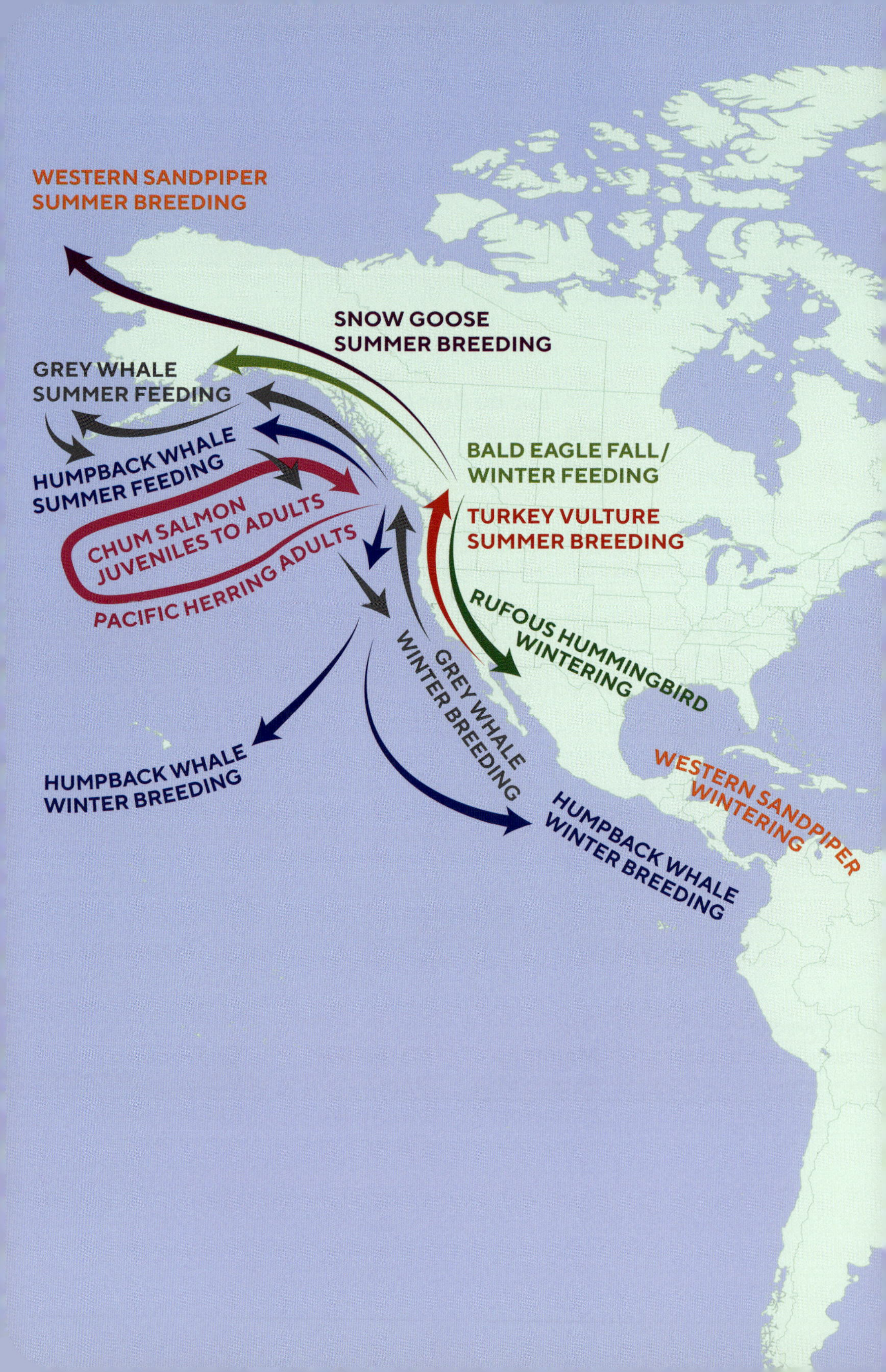
WESTERN SANDPIPER
SUMMER BREEDING
SNOW GOOSE
SUMMER BREEDING
GREY WHALE
SUMMER FEEDING
HUMPBACK WHALE
SUMMER FEEDING
BALD EAGLE FALL/
WINTER FEEDING
CHUM SALMON
JUVENILES TO ADULTS
PACIFIC HERRING ADULTS
TURKEY VULTURE
SUMMER BREEDING
RUFOUS HUMMINGBIRD
WINTERING
GREY WHALE
WINTER BREEDING
HUMPBACK WHALE
WINTER BREEDING
WESTERN SANDPIPER
WINTERING
HUMPBACK WHALE
WINTER BREEDING

Many species in this book have migrated vast distances to and from the sites you view them at. This map provides a general overview of some of these species' other critical habitat areas.

WITH PATIENCE, ONE CAN WATCH THE STEALTHY HUNTING STRATEGY OF A GREAT BLUE HERON AND, IF LUCKY, OBSERVE THE LASER SPEED OF A SUCCESSFUL CATCH.
PHOTO ROB ALEXANDER

CURIOUS ABOUT WILDLIFE?

I'd like to see the *Southern Resident Killer Whales,* but I am concerned about all the boat traffic around them. Can I watch them from land?

Yes, you certainly can watch them from land. Have you heard of the Whale Trail? One of the 22 sites on the trail is at East Point on Saturna Island. That is a great place for these whales and other marine mammal viewing at the right time of year. **See page 269. ›**

I know of a *Great Blue Heron* nesting colony that is right next to a Bald Eagle nest. Is there a reason why the herons are nesting near the eagles?

Good observation. Yes, it appears the herons nest near Bald Eagles because the nesting eagles are territorial, keeping other predatory birds away from the herons. The downside is the eagle pair see heron chicks as food for their own young. Researchers call it Mafia Protection! **See page 94. ›**

How many species of *hummingbirds* live in BC? And are they increasing or decreasing?

There are seven species of hummingbirds that come to BC, but only four that breed here. Some are increasing, and others decreasing. Anna's Hummingbirds are increasing, especially in winter, while Rufous are decreasing – there's worry about pesticides in their environment. You might be interested in Project Feeder Watch. **See page 113. ›**

If I see a *Black Bear* on a trail, should I make myself look big?

No. That would be a challenge to the bear. Unless it's moving toward you aggressively (which is very rare), you should speak softly and calmly to it, without looking it directly in the eye, and back away slowly. See page 127. ›

What is the *most endangered mammal* in BC and is it possible to see it?

Typically, I wouldn't send people to view an endangered animal. But in my research for this book I learned the **Vancouver Island Marmot Recovery Foundation** *is encouraging people to view these animals at Mount Washington. People keep predators away.* See page 157. ›

I just completed the West Coast Trail and I really want to *learn more* about the natural history of the west coast. Do you have any ideas?

Absolutely. What a great place to learn about. Check out the chapter on Grey Whales and Marine Mammals; in it you will learn about **West Coast N.E.S.T.** *It's convened through the Clayoquot Biosphere Trust.* See page 73. ›

I don't see as many, or the variety, of *wild birds* on my property as I used to. Why is that?

In 2019 a report came out that said the population of birds in North America had dropped by three billion since 1970! There are many reasons for this. Consider the **Seven Simple Actions to Help Birds** *outlined in this book and read about the importance of insects.* See pages 138–139. ›

I spend a lot of time in nature. Is there a way *I can connect to a wider community* to share what I'm seeing?

Yes, you certainly can. Check out **iNaturalist** *and* **eBird**. *If you are watching whales, look into* **Happywhale**. *These are Citizen Science apps that help you share what you are seeing with the wider naturalist and scientific community. And don't forget your local BC Nature naturalist clubs.* See pages 185, 286, and 366. ›

ALPINE MEADOWS TYPICALLY FLOWER FROM EARLY JULY AS A FIRST BLOOM TO MID-AUGUST FOR A SECOND BLOOM. THEY ARE CRITICAL FORAGING AREAS FOR MANY SPECIES.

PHOTO ROY JANTZEN.

THE HUMPBACK WHALE NAMED HEATHER, ALONG WITH HER CALF, ARE COMMON VISITORS TO THE SOUTHERN GULF ISLANDS REGION. *WATCH FOR THEM OFF SATURNA ISLAND IN EARLY SUMMER.* PHOTO LUCY QUAYLE.

FOREWORD

BY PATRICIA THOMSON

Marvin, a solo gentleman in his 70s, was a participant in one of the innumerable western Canada tours that I've had the pleasure to guide over many years. His lifelong pursuit of wildlife watching was the reason he was onboard for a most memorable day of experiencing Pacific marine mammals off the Gulf Islands. He enjoyed the summer warmth and the gentle rock of the boat. He could smell seaweed on the turn of the tide and even the fishy breath of sea lions hauled out on exposed rocks. He could hear raucous gulls, the slap of Pigeon Guillemot wings on water takeoff, and diverse accents of eager international ecotourists. Marvin had macular degeneration, however, and his field of vision had waned to a small doughnut of sight. He couldn't see much of the vast silver sea, the green island horizons, nor anything bobbing between.

It was a calm, undulating hour into our adventure when suddenly a rippling of exclamations, wild gesticulations, and camera clattering heralded a Humpback launching from the strait at a distance. I scrambled to find Marvin, determined for him to have the chance to see it too. Yes, he agreed to let me guide his focus toward my best guess for the whale's next breach. We missed its next leap, readjusted, missed it again, readjusted... "I saw it! I SAW it!" His ecstatic repetition was so simple, yet etched a lifelong joy for him, which was joy for me!

My own early encounters with wildlife started as a 3-year-old watching a salamander shed and eat its outgrown outer skin. When I was 7, the whistles of Marmots and Pikas lured my tired legs up steep mountain switchbacks on family backpacking trips in the Coastal Ranges. When I was a pensive preteen, a Pacific Giant Octopus entranced me for hours within the cosmos of a Port Renfrew tide pool.

I remain indebted, however, to two amateur naturalists who scooped me under their wings when I secured my intro job with BC Parks in Terrace. They were the first to begin unlocking all of my senses to the songs and fragrances of the forest. I was their "Marvin" as they tuned me to the thrumming of a Ruffed Grouse, the jaunty song of a Ruby Crowned Kinglet, the heady fragrances of Skunk Cabbage and musty duff.

When I "temporarily" moved – now almost 30 years ago – from the forested north coast to Metro Vancouver, it was under my illusion that I was settling to work in a canyon of concrete, and that my evolution as a budding naturalist would become bogged down. How happily wrong I was! Fields and forests, mountains and rivers, estuaries, ocean and, indeed, bogs absolutely define the Lower Mainland and its island communities. Wildlife abounds here.

Whatever piques one's interest in creature connections seems to cause a common lifting of the spirit. Nature's grace, beauty, and power inspire art and poetry. Nature focuses photography, tells stories in real time, and enriches lives.

Studies show that human happiness increases in contexts of higher biodiversity: glimpses of a Coyote crossing a fairway, a Western Painted Turtle basking in a park pond, an Anna's Hummingbird visiting balcony feeders throughout the winter. How can one not be at least intrigued, if not mesmerized, by the overhead arteries of some 10,000 crows that migrate dawn and dusk between their congregate roosting site at Burnaby's Still Creek and distant feeding grounds? Social media stats surge when videos are uploaded of River Otters sliding on a frozen Lost Lagoon, of Northern Flying Squirrels gliding in the heights of Stanley Park, of a Beaver that waddles between totem poles and tourists, of Killer Whales cruising under the First Narrows Bridge. When a young buck swam from North Vancouver across the currents of Burrard Inlet and spent a number of months moving throughout the populous West End, "Downtown Deer" became a veritable news celebrity. And when Vancouver asked its citizens to choose a Bird of the Year, more people voted in that poll than in local political elections!

We have an innate sense of curiosity and yearn to experience deep and binding connections to everyday wonders. We want to feel roots in our history on Earth. We hope for the survivorship of fragile creatures and vulnerable species. We desire to know that these living beings continue to share our world. But there is a need to respectfully understand that it is we who share theirs. Since moving to Metro Vancouver, I am sad for the declines in many animal and bird populations, tolls taken by habitat loss, climate change, or other reasons. There is a need for us to steward and sustain these species.

It was while I was heading Continuing Education programs for the Vancouver Aquarium in the late 1980s that, as an eager volunteer, Roy Jantzen became my biology buddy. Eager to study every creature from sandworms

ONE OF THE RARER SMALL MAMMALS THAT INHABIT MANNING PARK IS THE GOLDEN-MANTLED GROUND SQUIRREL. WHILE MANNING PARK IS AT THE NORTHERN EDGE OF ITS RANGE, IT IS OFTEN SEEN AT THE CASCADE LOOKOUT.
PHOTO LARRY SYROISHKO.

to sandpipers, we guided busloads of participants on such weekend immersions as a Grey Whale Getaway in Ucluelet, Sea Otter Odyssey in Kyuquot, and Sea Lion Safaris off Steveston and Nanaimo.

In time, Roy transitioned from student to environmental education professor and from program participant to program designer-deliverer. A sought-after lecturer and author, Roy now has decades of experience in tour leading, from alpine to coast to desert environs. He has immersed himself, his family, and countless others in epic Sockeye Salmon and Snow Geese migration events and in seasonal celebrations of Meadowlarks, Brant's Geese, and Bald Eagles, to name a few.

It is no surprise – yet a great delight – that Roy has been compelled to craft this guide to wildlife viewing within British Columbia's lower backyard. He has taken great care to hone in on what, why, where, and – with sensitivity – how to seek wildlife within diverse landscapes and seasons. *Wildlife Weekends in Southern British Columbia* is a resource to make engagement with nature accessible, easy, and insightful for any age. Whether for hobby, entertainment, recreation, or education, stewardship is at its core.

This is a set of guiding hands to put us in right places at right times, focusing all of our senses for meaningful "Marvin" moments!

Patricia Thomson *is a lifelong naturalist and environmental educator. She has guided more than 50 remote rafting expeditions spanning BC to the Beaufort Sea. She has worked for National Geographic Expeditions, Parks Canada, BC Parks, Vancouver Aquarium, and the Stanley Park Ecology Society.*

TIDE POOLING TYPICALLY INVOLVES WALKING OVER ROUGH AND SLIPPERY AREAS CLOSE TO THE OCEAN'S EDGE. *PAY CLOSE ATTENTION TO YOUR AND OTHERS' SAFETY WHEN EXPLORING IN REMOTE AREAS.*

INTRODUCTION

Wildlife Weekends: What Are You Doing This Weekend?

So, what *are* you doing this weekend? How often do we ask, or are we asked, this question? Our minds may go in multiple directions – Relaxation: "I'm going to kick back and read my book"; Social: "I'm going to catch up with friends or spend time with family"; Physical: "I'm going to get outside in this beautiful weather." Inevitably these answers lead to plans and plans then lead to experiences. *This guidebook is about experiences* – thoughtful, ethical, and edifying wildlife experiences that place you closer to nature, but without imposing on it.

Wildlife Weekends has been written to take the reader on a journey, both physically and metaphorically. Physically, it leads the reader to have outstanding experiences in beautiful southern British Columbia locations. Metaphorically, I'm hoping it evokes curiosity in the reader, inspiring them to ask deeper questions, to come to novel understandings, and to experience a sense of wonder, discovery, and ultimately of awe – a spiritual connection to the natural world.

Watching wildlife is not just about the end goal of viewing a particular animal, it's also about the journey getting there. The natural world is beautiful and complex. The skies dance with colour and shapes. I think of it like the Super Mario Bros. video game I played as a child, in which I gathered points by interacting with "tokens and jewels" along the route. So, as you proceed to your wildlife destination, pay special attention and be on the lookout for these "jewels." They may be in the form of a nibbled squirrel cone on a trailside stump, the high-pitched ascending song and subsequent

AMONG THE MOST COMMON FROGS HEARD AROUND METRO VANCOUVER IS THE PACIFIC TREE FROG. ***LISTEN FOR THEM NEAR FRESHWATER BODIES IN EARLY SPRING.***
PHOTO ROB ALEXANDER.

octave drop of a Golden-crowned Kinglet in the canopy, a trailside earthy odour that makes you stop and assess its source, or the dappled light of sun and shade that bathes the area as you cross soft spongy soil. These are the jewels of the journey, and they are not to be missed as you attend to your wider purpose of viewing your eagle or bear, butterfly or bat.

My name is Roy Jantzen and I am an environmental educator, naturalist, and author of *Active Vancouver: A Year-Round Guide to Outdoor Recreation in the City's Natural Environments* (Rocky Mountain Books, 2015). I have a Master of Arts in Environmental Education and Communication from Royal Roads University and have had a career as an educator at Capilano University. *Wildlife Weekends* is an invitation to discover the wildlife that inhabit the ecosystems of southern British Columbia. My hope is that greater *awareness* may lead a reader to form a deeper *appreciation* and *understanding* of animals in their habitats, ultimately resulting in more people placing greater value on our role in *protecting, conserving,* and even *rewilding* ecosystems. Generally speaking, society has become increasingly out of touch with the natural world as urban people spend more time in environments of our own making, following schedules of our own design. Our home, work, and often even our recreational environments separate us from the rhythms of the natural world.

Wildlife Viewing: Our Responsibility

All of the animals profiled in this guidebook are making their best attempts to live full lives in their human-altered habitats. Their territories likely include concrete, invasive species, excessive noise, human waste, or introduced chemicals. The surrounding land may have been flooded, drained, diked, or farmed, and each alteration triggers a change that requires the animal to adapt. Often that means moving to another habitat that lacks the critical elements needed for survival, exposes them to increased predation, or is already the territory of a competing species. These less-than-ideal habitats lead to declining populations. As naturalist Aldo Leopold reminded us, "To keep every cog and wheel is the first precaution of intelligent tinkering." That is, each species is *necessary* to the healthy functioning of an ecosystem.

In the context of this guidebook, it's important to begin by reflecting on how we observe wildlife. Wild animals have a tight energy budget that correlates closely to the activities of the season, and it's important for a wildlife

SEA LIONS ARE KNOWN AS EARED SEALS, FOR REASONS OBVIOUS IN THIS PHOTO OF A MOM AND HER PUP. PHOTO MICHAEL SCHMIDT PHOTOGRAPHY.

observer to consider these. In *spring*, reproduction and migration trigger high energy demands. *Summer* is about finding enough food to feed their young. The main concern in late summer and *autumn* is putting on enough fat to survive the harsh weather of *winter* with its corresponding reduction in food supplies. The choices we make as observers have direct impacts on wildlife. For example, if a species knows you are in its vicinity, it may freeze to listen and observe, or to scold you with its call. This in turn may lead to an increase in heart rate, a loss of valuable foraging time just before nightfall, or a decision to flee, burning up additional energy. As I learned in my research, wildlife are encountering people all the time. To quote Todd Windle, a project manager I interviewed about Grey Wolves in Pacific Rim National Park, "To have an interaction with a wolf is pretty powerful. Every person calls it a *once-in-a-lifetime experience*. They don't realize that the wolf had that *once-in-a-lifetime experience* earlier that day, and will have another *once-in-a-lifetime experience* later that day, and again the next day, and then five more *once-in-a-lifetime experiences* the next week." It is often well-meaning people who love wildlife who can do the most detrimental things. Therefore, a wildlife-friendly lens will lead us to consider how our presence might affect the wildlife, and result in us making choices to their benefit, not ours.

On our planet, there is an ongoing and unnatural loss of biodiversity that is accelerating. This not only tears at the intricate fabric of our global ecosystem but also sabotages nature's power to heal, nurture, and replenish. The loss of, or even a substantial reduction in, a species population can have a domino effect, sending other species into a downward spiral and ultimately threatening an entire ecosystem. Consider the game Jenga where players can take out blocks from a tower and nothing happens until too many blocks are missing for the tower to remain standing. As populations of species diminish, the function each played in the system also vanishes. Some call these functions "ecosystem services" (outlined further on page 226), and our collective actions often compromise these free contributions to our lives, which species perform to our benefit.

Protecting and Coexisting with Wildlife

There are many ways we can protect and coexist with wildlife. The deliberate choices we make around our homes or when recreating in nature can attract or reduce wildlife interactions we *don't* want. A few of these deliberate actions include:

→ Secure compost or garbage bins and putting them at curbside to be removed *as close to the time of pickup* as possible.

→ Ensure our pets (especially cats) are not able to roam freely and hunt, while also safeguarding wildlife from pet food and other attractants by keeping them *inside* the home.

→ Remove dropped fruit (or unpicked fruit) from your yard to make certain it doesn't attract wild animals. There may be local food security organizations that will gladly pick fruit to share with others.

→ Use methods available to ensure windows and glass on decks don't become death zones for birds expecting to fly through them. To understand our impact on birds more thoroughly, I recommend 3billionbirds.org, and for solutions, the Cornell Lab's Seven Simple Actions to Help Birds.

→ If you do encounter injured wildlife, contact your local wildlife rescue association.

→ When recreating, don't leave food waste (eggshells, orange peels, apple cores, not to mention wrappers!) at snack or lunch stops, and don't give in to the Canada Jays or Ravens "asking" to share your food.

Wildlife Weekends Design

The "*Weekends*" design of the book is about wildlife viewing opportunities that are *no farther than a six-hour drive* from Vancouver, and that can be experienced in a Friday-to-Monday long weekend. I chose "*hardened*" wildlife viewing sites, meaning sites that have viewing infrastructure already in place, including trails, boardwalks, viewing lookouts and platforms, parking, and possibly even a visitor centre with restrooms. In choosing the sites, it was critical to me that I did not invite people into places where they would compromise or negatively impact the species they had come to view. I tried to choose sites where the presence of wildlife is fairly predictable, and that are also great places to visit, even if the species-of-interest doesn't show up as hoped. All are scenic, and many have additional cultural or historical significance, offering a broader depth in the overall experience.

The guidebook leads the reader to 20 wildlife viewing sites, covering an array of species, including mammals, birds, reptiles, amphibians, fish, invertebrates, and molluscs, and provides interpretation of the ecology and natural history for each. Each chapter focuses on the intersection between a main species attraction, a designated location, and a key time (e.g., Turkey Vultures in East Sooke Regional Park in the third week of September), with the objective of providing the reader with a place-based narrative that will connect them to the species in the context of its ecosystem. In addition, each chapter includes directions and accessibility considerations, maps (if the site may be difficult to find), distances (if walking trails are involved), other locations where you might find the species, other natural, historical, or cultural interests near the site, and further resources to consider the topic in more detail. In many of the chapters there are boldface terms in the text. The reader may consult the glossary near the end of the book for more information about these terms.

TURKEY VULTURES GLIDE ON RISING AIR MASSES. WITH THEIR WINGSPAN STRETCHED OUT OVERHEAD, IT IS OFTEN EASY TO SEE THE DUAL TONES IN THEIR WINGS. PHOTO TERRY CARR.

IT IS UNLIKELY THAT A FAMILY TRIP THAT INCLUDES WATCHING WHALES FROM SHORE WILL EVER BE FORGOTTEN IN THE MINDS OF CHILDREN. PHOTO CHRISTINE GAIO.

Wildlife Viewing Tips and Etiquette

Wild animals are more astute at watching than you are, and because of this they will likely have their ears, eyes, and noses alert and scanning, even before you arrive. Their movements are calculated to keep them from being preyed upon, which means they try to go unnoticed. With some effort, however, you may be able to encounter the wildlife on their home turf, but remember: *effort doesn't guarantee success.* Wild animals are unpredictable, so although you may arrive at the "right" time and place to see a species that is habitually there, they still might not show up. When they do, however, the ultimate wildlife viewing experience is behaviour watching, observing the animals without interrupting their normal daily activities. It's the difference between identifying an animal and being able to *identify with it*. Therefore, it is essential to consider the following observation tips to help improve your chance of success, and to master the "how" of wildlife viewing:

→ Make yourself as invisible as you can. Bright clothing or even the reflection off your glasses can alert an animal to your presence. Placing yourself

DON'T FORGET YOUR TOOLS – YOUR BINOCULARS, THAT IS – WHEN YOU HEAD OUT ON YOUR NEXT WILDLIFE VIEWING TRIP. PHOTO RONALD LAUBENSTEIN/USFWS, WIKIMEDIA COMMONS, PUBLIC DOMAIN.

behind vegetation, against a tree or against a boulder, will break up your outline and ensure you are not so obvious. Watching from a vehicle also helps by hiding your silhouette. If an animal sees you, avert your gaze as animals may interpret your stare as a threat and leave the area.

→ Wild animals don't just use their eyes; they sense through their noses, their ears, and their toes as well, so avoid strong scents, keep your voice down, and walk softly to minimize ground vibrations. Mammals will disappear (by freezing or fleeing) long before you arrive if you let them know you're coming, so using measured movements and decreasing scents and sounds is critical. Does a stiff breeze make you more noticeable because you are upwind, or does a calm evening make your sounds carry farther?

→ If you turn over leaves or logs in search of amphibians, do your best to replace them just as you found them. The same applies for rocks turned over in an intertidal area. You're lifting the roof of their home! Nests are especially sensitive to our touch, as we leave scent trails that may lead predators to a nest they might otherwise have overlooked, or cause mothers to abandon their babies because they no longer recognize their scent as their own.

- → Look up, down, and all around, including using your peripheral vision. Focus in one area for a few minutes and then scan your wider surroundings to gain a sense of context. Every few minutes, switch between focusing and expanding your view. Watch for out-of-place movements. Use as many of your senses as you can. Deepen your awareness by tapping your senses of sight, hearing, and smell. Owls move on branches above your head, woodpeckers call from tree holes, songbirds sit silently in nests at trailside, while mammals often urinate on trees or browse the leaves and branches of the waist-high shrubs in their territories. Make "mule ears" by cupping your hands around the backs of your ears to amplify natural sounds.

- → Do your research before going out. Where does the animal you are looking for spend its time? What is its preferred habitat or niche? Where is the activity you hope to observe going to occur? Are the twilight hours of sunrise and sunset the best time to look for it? Can the light of a full moon or dark of a new moon be helpful? What about high or low tides? Questions like these allow you to concentrate your efforts and increase your likelihood of success.

- → Use aids such as a camera with a zoom lens or a headlamp with a red light, as many animals don't see red. Good quality binoculars (7x35 or 8x40 with central focusing) are also useful. Several Metro Vancouver libraries even allow you to borrow binoculars (and night-sky telescopes too) with a library card. The binoculars come with a kit that includes instructions, as well as a birding guide. Kits can usually be borrowed for 14 days at a time.

- → Download the **Seek** app, by iNaturalist, to help you identify species you see.

- → Ask local naturalists for viewing advice, as locals are often very familiar with viewing opportunities and locations for the species in their areas.

- → Always consider your own personal safety. Bring proper clothing and footwear, extra food and water, first aid, and a flashlight or headlamp along on your viewing experience, and make sure you have automobile safety equipment in your car. In addition, pay attention to hunting seasons. Most fall between September and November, but some openings continue into spring, or are year-round.

→ Please observe the following guidelines when viewing wildlife:

- Respect the animals, the land, and other wildlife watchers. The Outdoor Recreation Council of BC promotes trail ethics with the "**Three C's**": Common Sense, Communication, and Courtesy.
- Know the fine line between observing wildlife and interfering with wildlife. If you can see you're affecting an animal, back away.
- Respect the laws of private property.
- Do not feed wildlife in natural areas: handouts of human food do wild animals no favours.
- When in a BC park, know the Responsible Recreation guidelines they ask users to follow.
- Stay on established trails, leave no trace, ensuring the habitat has not been altered due to your presence there, and be extra vigilant in sensitive areas such as wetlands and highly vegetated areas with thin soils, such as when walking in alpine ecosystems.

Wildlife Viewing as a Family Affair and for All Abilities

Do you have a child or teen who shows an interest in birds, insects, reptiles, or other wild creatures? Do they have an affinity for nature, science, or the outdoors? Encouraging and nurturing these pursuits is what the planet and its citizens need right now. Please use the wildlife viewing chapters in this guide (along with the appendix on page 407 dedicated specifically to wildlife viewing experiences with children and youth) as an educational resource to help encourage their curiosities.

Do you know others who love nature but often find barriers to their participation? I researched wildlife viewing sites with accessibility options in mind and have attempted to share what I found in each chapter's Directions and Accessibility section. Audubon, for example, offers a North American Birdability Map that "works to ensure the birding community and the outdoors are welcoming, inclusive, safe and accessible for everybody." Locally, the Great Blue Heron Nature Reserve in Chilliwack has wheelchair-accessible buildings and trails for birders of all abilities.

Engaging and Taking Action for Wildlife

How do you engage or take action to protect endangered wildlife and their habitats? Engagement is critical and many BC species need our attention. A prime example is the Deep-Snow Mountain Caribou, whose southwestern

A BRITISH COLUMBIA MOBILITY OPPORTUNITIES SOCIETY (BCMOS) TRAILRIDER IS USED AT A BRIDGE CROSSING TO ENABLE AN ADAPTIVE RECREATIONALIST ACCOMPANIED BY BCMOS VOLUNTEERS TO HAVE A BACKCOUNTRY EXPERIENCE. PHOTO BCMOS.

BC habitat includes the Wells Gray region. Their numbers are so precarious that I cannot guide a reader to a place they are *likely* to encounter them. In 2005, there were 18 herds. Today, only ten herds remain and, of those, only six are considered viable. In 2019, the two southernmost herds became locally extinct. In 2016, all of the remaining herds totalled barely 1,000 individuals, and their numbers continue to decline. Their range includes the Wells Gray region, the Cariboo Mountains, the Hart Ranges, and possibly the Central Selkirk Mountains. These animals exist nowhere else on Earth. They are moving toward extinction and they are moving there fast. Some of the reasons for their precariousness are outlined in the chapter on Moose in Wells Gray Provincial Park. So far there are not enough outraged, engaged, proactive, and vocal British Columbians to keep the BC government from making bad decisions that have compromised the Caribou and their habitat. You can bear witness to the decline of this species at the scientifically credible website 1000clearcuts.ca.

This book is about the joy of viewing wildlife, but it should also be a reminder of what wildlife species you can't view anymore, *because they don't exist anymore*. We are in the midst of a sixth global extinction – a

SOUTHERN MOUNTAIN CARIBOU IN BC ARE ONE OF THE MOST ENDANGERED LARGE MAMMALS OF THE PROVINCE. IT'S WORTH LEARNING MORE ABOUT THE CONDITIONS THAT PUT THEM AT RISK. THEY ARE DISCUSSED BRIEFLY IN THE CHAPTER ON MOOSE.
PHOTO STEVE FORREST, "WOODLAND CARIBOU," FLICKR, CC BY-NC 2.0.

biodiversity emergency! So many species need us to understand their predicaments now. Although Deep-Snow Mountain Caribou are an iconic species (they're on our quarter!), BC also has many other species that don't exist any longer and therefore cannot be part of any wildlife viewing book. From Basking Sharks to Northern Abalone, from Northern Spotted Owls to Oregon Spotted Frogs, continued existence for many BC species is uncertain. In every chapter of this book, there are resources to learn more. *Please learn more.* There are suggestions for ways to engage and take action around wildlife issues. *Please engage and take action.* As citizens, we must take on a level of responsibility that matches the gravity of the changes BC's species are experiencing. We mustn't rely only on others to do the learning and the work.

To address this, within each chapter under the heading *Engaging and Taking Action,* I have discussed human-caused impacts on wildlife and included ideas for clear actions we all can take to mitigate or remedy these impacts. If upon reading a chapter you have been moved to action, ask yourself how you, personally, can contribute to *positively affecting* the wildlife and the ecosystems you plan to view. Here is a spectrum of measures you might consider, including:

→ **Simple Actions/Commitments:** Put out a feeder for local songbirds or hummingbirds, or commit to learning more, through this guidebook or through the online BC Nature Guide, which covers nature sites, naturalist clubs, and events throughout BC. Start by viewing the interactive map and see where it takes you.

→ **Moderate Actions/Commitments:** Join a local naturalist club or participate in a "citizen science" project like Birds Canada's Project Feeder Watch. Take a course from the Rocky Point Bird Observatory or through WildResearch, and then volunteer on a local bird banding team.

→ **Deep Actions/Commitments:** Commit time, effort, and life experience to a conservation organization. To do so is a virtuous act. There are many needs, but too often volunteers, once trained, are transitory. Is there an environmental organization you believe in where you think you can make a difference? See the *List of Referenced Resources & Organizations* outlined in this book. What are you waiting for? There is no time like the present.

→ **Career-oriented Actions/Commitments:** Start on or revise a career path toward environmental research, conservation, or habitat restoration.

BANDING HUMMINGBIRDS AND ANALYZING HUMMINGBIRD URINE HELPS SCIENTISTS UNDERSTAND HOW THEY ARE RESPONDING TO CHEMICALS IN THEIR ENVIRONMENT, SUCH AS THOSE SPRAYED ON BLUEBERRY FIELDS. PHOTO ALISON MORAN.

There is also much more to learn by visiting the BC government and nongovernmental organization websites and using the search term "endangered, extirpated, and extinct species in British Columbia."

Please send me a note that shares how you have engaged and taken action, or send ideas and available opportunities. I will share them widely.

Minimizing Your Carbon Footprint

Every chapter in this guidebook includes ideas for *minimizing your carbon footprint*. This was a central objective when envisioning this project because, in my opinion, we mustn't ignore OUR impact on the species with whom we share the Earth. The economic, social, and environmental risks of unabated, **greenhouse gas**–caused climate change are immense, and our fossil-fuel-based energy usage is at the heart of this. We use energy to keep us comfortable, to transport us, and to allow us to work long after the sun goes down, but with over eight billion people on the planet, we are kidding ourselves if we think other species are unaffected, or that it can continue. I read the journal *Nature* to help me understand the latest developments in the world of science. Several years ago I saved a *Nature* editorial quote that reads, "Take a step back. Earth is an orb reeling through the emptiness of space, warmed from the inside by molten rock and from the outside by the Sun. Thanks to the wonders of physics, a magnetic shield protects us from the most damaging cosmic radiation, and our exceedingly thin *atmosphere forms a protective blanket* that keeps the surface temperature at more or less acceptable levels for a wide variety of life. Humans have no control over the former, but it's now crystal clear that maintenance of the latter [our atmosphere] *is on us*."[1] It is *on us* to mitigate the damage.

This critical realization meant I couldn't write a book and ignore the planet-heating carbon emissions involved with encouraging people to travel for a wildlife viewing experience. The direct and indirect effects on wildlife are embedded in the decisions we make. Flood this, dam that, burn this, capture that, build here, clear there. But not all decisions are equal, and some have more substantial impacts on wildlife than others. The idea of *minimizing your carbon footprint* is included as a challenge to think differently, and as an acknowledgement that we are not only the species that has the greatest impact on wildlife, we are also the one with the greatest capacity to protect, repair, and, ultimately, improve the way we coexist.

Choosing a *"carbon-light"* mode of transport when planning a trip to view wildlife will likely take you longer; it could see you travelling with others you don't know; or it could have you burning more calories getting to the location than if you drove. While this guidebook offers suggestions

NORTHERN HARRIERS ARE COMMON ALONG THE DIKES AROUND BOUNDARY BAY AND WESTHAM ISLAND. LOOK FOR THEM FLYING LOW OVER THE GROUND AS THEY SEARCH FOR MICE AND VOLES. PHOTO SABINE DECAMP.

to *minimize your carbon footprint* for all chapters, some are easier to consider than others. Here are some of those suggestions:

→ **Transit or transit-and-bike** options are suggested for Metro Vancouver and Fraser Valley locations such as Great Blue Herons, Winter Birds of Metro Vancouver, Coyotes, or Salmon at Goldstream Provincial Park.

→ **Bus companies** offer service between Vancouver, Vancouver Island, and the interior of BC, allowing the public a carbon-light option to get to viewing locations such as Princeton for hummingbirds, Peachland for bats, or Courtenay-Comox as the launching point for Mount Washington's Vancouver Island Marmots.

→ **Cycling** is suggested for sites that have good infrastructure for cycling (such as the Lochside and Galloping Goose trails to access East Sooke Regional Park and the Turkey Vulture migration), for "near-local" areas (such as the Fraser Valley), and for locations where a car and bike could be combined (such as the Meadowlark Nature Festival in the South Okanagan).

→ **Electric car or e-bike rentals, or car sharing** options (Evo or Modo) are suggested for wildlife viewing sites that are within a day trip of Vancouver (including Bald Eagles in Brackendale or dragonflies near Hope).

→ **Filling a vehicle with family or friends** decreases per person emissions on a trip and works well for certain experiences, such as tide pooling at Botanical Beach, viewing Black Bears in Wells Gray Park, or watching Killer Whales and other marine mammals from Saturna Island.

In BC, electricity to power electric vehicles mainly comes from hydroelectricity, not from fossil fuels, but everything comes at a cost. Hydroelectricity does require the logging and flooding of wildlife habitat, the building of dams for electricity generation, and the stringing of thousands of kilometres of power lines through tree-free corridors to transport the electricity. It's not impact-free, but it is less impactful than a fossil fuel alternative.

Writing Contributors to *Wildlife Weekends*

I attribute the development of my environmental education knowledge and mindset to some very special people over my lifetime. In this project I reached out to five of them to ask if they would contribute to this book. Four were asked to provide an introduction to a season, and the other to write the foreword to the book. I was ecstatic when they each said yes! These contributors are:

Ornithologist **Dr. Robert Butler** wrote the *Introduction to Spring*. His book *The Jade Coast: The Ecology of the North Pacific Ocean* really helped me understand the Salish Sea better.

Dr. Richard (Rick) Kool wrote the *Introduction to Summer*. Rick is a lifelong educator and was responsible for much of my learning in the Master of Arts in Environmental Education and Communication program, which he managed at Royal Roads University.

CBC meteorologist **Johanna Wagstaffe** is a media science reporter, and I find I am always watching and listening for her interpretation of the latest science news. She wrote the *Introduction to Autumn*.

Although I had never met **Anne Murray** prior to this project, I was in awe of her volunteer service, often reading what she was currently involved in through BC Nature, an umbrella organization of BC naturalist clubs. Anne is a great example of someone who serves her community and the wildlife that inhabits it. I asked her to write the *Introduction to Winter.*

→ There really was only one person I felt who could write the *Foreword* to this book, and that was my friend of 30-plus years, Patricia Thomson. I first met Patricia when I worked as a volunteer with the Vancouver Aquarium education programs. We stayed connected over the years – she worked as a rafting guide and I'd always be there waiting for her to share her stories after returning from northern rivers like the Tatshenshini-Alsek. Most recently she was the executive director of the Stanley Park Ecology Society, a local conservation organization for which I became a volunteer myself.

Thank you for your interest in wildlife and in this book.

Please let me know how you have used this book and what you like about it...

Roy Jantzen
wildlifeweekendsbc@gmail.com

SPRING

THE VARIED THRUSH IS SEEN YEAR-ROUND IN SOUTHWESTERN BC. *LISTEN FOR THEIR SINGLE PITCH DIMINISHING FLUTELIKE SONG ANY TIME OF THE YEAR, BUT ESPECIALLY IN THE SPRING.*
PHOTO ERIC GROSS.

INTRODUCTION TO SPRING

The Green and Silver Waves

BY DR. ROBERT BUTLER

Ornithologist, artist, and conservationist. Author of *Pacific Flyway* and *Jade Coast*. President of the Pacific Wildlife Foundation, Fellow of the Explorers Club, and Royal Canadian Geographic Society. Promoter of the Nature Culture concept. See robbutler.ca/.

Over 4.5 billion years ago, a giant asteroid collided with a newly formed Earth, knocking it a little off kilter. The planet would never be the same. From that moment on, the great wobble resulting from that chance collision gave rise to our seasons, enhancing conditions on Earth required for the evolution of life. For half the year, the northern hemisphere tilts away from the Sun so that sunlight strikes the Earth at a low angle. Less direct sunlight creates colder conditions, characterizing our northern winter. For the remainder of the year, the northern hemisphere tilts toward the Sun. The higher angle of the Sun brings more direct sunshine, and this, in turn, creates the warmer weather characteristic of our summers. Spring is the period when the tilt of the Earth transitions between cool winter and warm summer.

As it does in all temperate regions of the world, springtime in British Columbia brings longer and warmer days. Low pressure weather systems in the Gulf of Alaska dominate in winter as they draw prevailing winds along our coast from the southwest, the winds that hurl rain onto our shores. As spring approaches, high pressure weather systems get the upper hand. The rain clouds become less frequent so that the Sun shines more often, and the winds reverse direction and arise more frequently out of the northwest.

As a young biologist living on a remote coastal island, I witnessed the sparring of high and low pressure systems. Southeast gales that howled for days, slapping branches against my cabin and tossing waves onto the beach, suddenly stopped, and in the stillness the clouds parted to reveal stars overhead. I was in the eye of the storm. Before long a faint breeze rose out of the northwest, and within minutes the island was once again assaulted by wind and waves. At the time, the winds were a deterrent to travelling by boat, but I was to learn that they were also the engine of life.

These Pacific low pressure and high pressure systems, which dictate the direction winds blow, also play a role in ocean currents, and ultimately in the ecology of our coast. Beginning about April, increasingly prevalent northwest winds rushing out of the Pacific high pressure over the Pacific Ocean, along with the spinning force of the Earth's orbit known as the Coriolis Effect, nudge surface seawater offshore. In response, cool nutrient-rich waters from the deep ocean abyss rise to fill the void, bringing their nutrients to the surface and into the sunshine. In this nutrient-rich soup blooms a meadow of phytoplankton, the foundation upon which all other coastal marine life is built, and so vast it is visible from space! The increasing sunlight as the Earth's northern hemisphere tilts further toward the Sun during spring triggers all green plants and marine algae to re-engage in the greatest ecological phenomenon on Earth, the process of photosynthesis. Both in the ocean and on land, the choreography of the grand dance of life, *the green wave,* has begun.

The Green Wave

At the southern edge of British Columbia, some plants begin to sprout new leaves as early as January, and with each successive day the landscape sheds its winter tawniness for a fresh new verdant green coat. This "green wave" of sprouting plants edges northward, moving more quickly along the mild coastal regions and valley bottoms than through the cooler interior or mountain slopes. These newly emerging plants are changing too. The nutrient content of the leaves goes from the succulence of new birth to a coarseness with maturity, so that within the green wave there is also a *nutrient wave,* not readily apparent to our eyes, but not unnoticed by plant-eating insects and birds. Arctic nesting geese, anticipating a long migratory flight to Arctic nesting grounds ahead of them, appear to exploit the nutrient peak of the green wave far to the south of their breeding grounds by nibbling the tender young shoots early on. They then depart northward, overtaking the green wave to match their arrival on the breeding grounds with the arrival of spring. As April arrives, Snow Geese in the great Fraser River delta marshes can be seen "high-grading" the young green shoots of sedges and rushes.

Far to the south, in Mexico and Central America, the songbirds and shorebirds sense the urge to migrate. By April, young leaves and flowers on our local trees and shrubs attract insects in abundance for the arrival of songbirds! As if appearing from nowhere, warblers, vireos, flycatchers, tanagers, and grosbeaks seem suddenly to be everywhere in our coastal forests, and a few weeks later they appear in interior forests as well. Insect larvae also migrate or emerge to nibble tender leaves and drink plant sap, which draws flocks of migratory insectivorous songbirds. Some insects migrate north in mind-boggling numbers. One estimate from the United Kingdom puts the biomass of 3.5 trillion insects arriving in spring at 3200 tonnes!

During the pandemic year of 2020, I stopped by a neighbourhood park following a wet cold spell in early May. There I found *parties* of freshly coloured Wilson's Warblers, Yellow-rumped Warblers, and Black-throated Gray Warblers spending the day darting to and fro in the deciduous forest in pursuit of food to fuel the next leg of their journey, which would take place later that evening. Similar scenes were playing out across my entire neighbourhood, wherever there were suitable trees. The cool weather had stalled the migration, which allowed me a few hours to marvel at the magnitude of the spring passage of songbirds.

Along the shores of the Fraser River delta, another tsunami of birds arrives in late April, when hundreds of thousands of Western Sandpipers arrive from their winter quarters in Central America. They, too, are seeking food to refuel for the flights that will carry them into Alaska where, ultimately, they'll nest. The sandpipers feed on "biofilm," a collection of microscopic organisms that grow together on the surface of the mud, to help build up the fat reserves required to fuel their journey. But they also have another trick – one that originates in the winds. The same low pressure systems that track onto the BC coast every few days throughout spring bring winds from the south that rush northward, and this is precisely the direction of the sandpipers' migration! By timing their departure with the winds, sandpipers get a boost to their flights with tail winds at high elevations to jet them across the sky on a trajectory that will not end until they reach their breeding grounds in western Alaska.

The Silver Wave

Birds are not the only animals that prepare for breeding as spring approaches. While winter still dominates the North Pacific, herring prepare to spawn along coastal British Columbia. Immense schools move closer to shore, and in late February and March the orgy begins. Males release milt (sperm) into the water, creating underwater milky clouds for kilometres of shoreline. Females, laden with eggs, spew their little gem-like ovules onto shallow beaches amid the sperm. Over the next few weeks, trillions of tiny silver herring larvae will emerge, grow, and, if they are not fed upon by another organism, they'll move to coastal areas to take advantage of the profusion of plankton.

Wherever there is an abundance of eggs and fish, another suite of predators is sure to follow. The "silver wave" of spawning herring that moves northward along the coast in spring draws Grey Whales migrating from their Mexican breeding grounds to the British Columbia and Alaska coasts to gulp down herring eggs. The herring also draw Surf Scoters in the thousands from local and distant beaches to attend the feast. These sea ducks follow the progression of herring spawning northward into Alaska. The herring eggs attract many different bird species, including loons, grebes, gulls, and an array of other sea ducks. Mammals join the frenzy too, with California Sea Lions and Harbour Seals swimming after the fish bonanza at sea, and wolves, bears, and mink joining in on the beaches. Some of the best places to see this annual spectacle of nature are the Vancouver Island beaches near Hornby and Denman islands (see the chapter on Herring in the Strait of Georgia). You will likely hear sea lions barking from offshore, and if you arrive at the peak time, you might see 20,000 gulls, along with thousands of other birds, feeding on herring or digesting their meals along the shores of the islands.

Although more than four billion springs have passed over the Earth, far outstripping our own evolutionary time on the planet, to us each spring brings a renewed sense of wonder, as if each year were a completely new phenomenon.

References

Hu, G., K.S. Lim, N. Horvitz, S.J. Clark, D.R. Reynolds, N. Sapir, and J.W. Chapman. "Mass seasonal bioflows of high-flying insect migrants." *Science* - AAAS 354, no. 6319 (2016): 1584–87.

Kuwae, T., P.G. Beninger, P. Decottignies, K. Mathot, D.R. Lund, and R.W. Elner. "Biofilm Grazing in a Higher Vertebrate: The Western Sandpiper, Calidris Mauri." *Ecology* 89 (2008): 599–606.

Lok, Erika K., Daniel Esler, John Y. Takekawa, Susan W. De La Cruz, W. Sean Boyd, David R. Nysewander, Joseph R. Evenson, and David H. Ward. "Spatiotemporal Associations Between Pacific Herring Spawn and Surf Scoter Spring Migration: Evaluating a 'Silver Wave' Hypothesis." *Marine Ecology Progress Series* 457 (2012): 139–50.

Yali, S., Q. Xin, W.F. de Boer, P. Gong, R.C. Ydenberg, and H. Prinz. "Do Arctic Breeding Geese Track or Overtake a Green Wave during Spring Migration?" *Scientific Reports* 5 (2015): Art. 8749.

EGGS OF HERRING – KNOWN AS ROE – ARE OFTEN CAPTURED ON INTERTIDAL SEAWEED. ***CONSIDER THE CULTURAL SIGNIFICANCE OF GREAT SPAWNS OF HERRING ROE TO FIRST NATIONS HUNTER-GATHERERS.***
PHOTO REBECCA BENJAMIN-CAREY.

Pacific Herring

in the Strait of Georgia & the Hornby Island HerringFest

AT A GLANCE

WHAT: Pacific Herring (*Clupea pallasii*)
WHERE: HerringFest on Hornby Island
WHEN: Early March
TRADITIONAL TERRITORY: K'ómoks, Homalco, Wei Wai Kum, We Wai Kai, ɬaʔamɩn gɩj̓ɛ (Tla'amin/Sliammon), Coast Salish
ACCESS: Easy. May also require time on a boat.

For thousands of years, and *hundreds of generations*, Pacific Herring have arrived on the BC coast to spawn, signalling the beginning of spring and of a new bountiful year. The Saanich, or W̱SÁNEĆ people, of southern Vancouver Island and the Gulf and San Juan islands relied heavily on this spawn. They called the full moon at this time the "WEXES" (the moon of the frog that announced the sacred season of plenty). The moon was a sign to lay Red Cedar boughs (also hemlock and kelp) in intertidal areas to provide protected sites for female herring to lay their eggs. Today, managers and scientists call herring a **forage fish** – dry terminology that really doesn't evoke the poetry of a species that is the cornerstone of ecosystems, with ripple effects to other organisms that are either directly or indirectly connected to them. When fish gather in one place by the thousands, millions, or even billions, in order to begin their reproductive cycle by laying and fertilizing their eggs, it's known as a fish "spawn." Archeological excavations in the Strait of Georgia and Puget Sound document dozens of archeological sites in coastal inlets and bays that show evidence of consistent herring spawn for thousands of years.[2] The herring spawn was, and still is, among the most important natural events on BC's shorelines, and it is central to the social, cultural, and economic relations of coastal Indigenous communities.

THESE ARE THE LAYERED REMAINS OF AN INDIGENOUS VILLAGE SITE ON THE BC COAST. THE LAYERS REPRESENT COUNTLESS GENERATIONS OF PEOPLE HARVESTING RESOURCES FROM THE SEA, INCLUDING HERRING, SALMON, AND SHELLFISH. PEOPLE AT THIS LOCATION WERE SUSTAINABLY CATCHING AND EATING HERRING FOR OVER 7,000 YEARS. PHOTO BRIAN PEGG.

If you travel in early March to Hornby Island's HerringFest led by Conservancy Hornby Island, you can expect two things. The first is a spectacular transformation of land and sea, where the wildlife of the marine, terrestrial, and sky environments come together to feast on the sudden influx of energy and nutrients the annual herring spawn brings to the region's sub-tidal, intertidal, and shoreline ecosystems. The second detail you'll notice is the presence of First Nations, commercial fishers, environmental groups, ecotourism operators, and residents, all of whom have deep connections to this species and their habitat. Many think overfishing practices have dangerously compromised this species-habitat system, and advocate for a moratorium on the commercial fishery, maintaining we have already taken too much. But there is disagreement here, and Fisheries and Oceans Canada (DFO) counter that they already take a **precautionary principle** approach, ensuring only a portion of the run is commercially fished. I interviewed Scott Wallace, a senior research scientist and marine ecosystem specialist at the David Suzuki Foundation about this. Scott told me that we're always looking for simple explanations to help us understand complex systems, but that can lead us astray in managing a species. But DFO knows, and Scott agrees, that the BC coast herring spawn is the single most important event, when considering the flow of biomass.[3] And important events bring out big emotions. Think of aerial videos you may have seen of the Caribou migrations in Canada's far north or the Wildebeest migrations in the Serengeti in West Africa to grasp some idea of what billions of fish concentrating themselves into a really small area, for just a few days, looks like. Herring are an indicator of a healthy ecosystem, even though we can't always predict their movements.

Today, the DFO conducts annual assessments of the herring, which incorporate historical data, commercial catch numbers, biological information, and spawn survey data. This helps them estimate current herring amounts and make predictions for up to a year into the future. Herring are largely governed by complex and uncertain processes, and naturally experience patterns of decline and recovery. While taking into account these uncertainties, the DFO currently allows fishers to take 20 per cent of the estimated herring stock. Scott Wallace of the David Suzuki Foundation believes that our management of herring is science-based and thoughtful, but he also knows there is room for error. He adds that this is because the estimated number of herring is extremely difficult to attain *before* seine and gill nets are already set in the water. It's important to get this right as so many other species also rely on this species. In 2021, he went on record, recommending the fishery be cut in half, to only 10 per cent of the estimated stocks, citing the need for the "utmost precaution," given herring's overwhelming importance in BC's coastal ecosystem.[4]

Historically, when the W̱SÁNEĆ harvested herring, they did it with a special rake, held by a fisher in a canoe who swept the rake back and forth through a school of herring, impaling the oily fish on the rake's tongs. What they took from the ocean, they used for themselves and for trading with other coastal and interior Nations. After 1873, though, once herring were seen as a valuable economic resource by the Canadian government and its precursor to DFO, things changed, and changed significantly. A historical

A BONAPARTE'S GULL MAKES A GREAT CATCH OF A HERRING MEAL. PHOTO SABINE DECAMP.

timeline of herring on the BC coast, produced by the Hakai Herring School, goes like this: By 1905 commercial boats were using drift gill nets, averaging 400–800 metres, scooping much larger quantities of herring than had been taken historically. By 1945, echo sounders and radio phones were used to access distant and deeper schools of fish. By the early 1960s, even more efficient fishing methods added sonars, drum seines, and mercury arc lights (to draw fish toward the boats). By 1968, coastal BC's herring fishery had collapsed, and the historic spawn and harvest sites of Point Grey, Nanaimo, Nanoose Bay, Fulford Harbour, Genoa Bay, Prevost Island, and countless other bays and shorelines along the coast of BC ceased to exist as they had in the past.[5] Local First Nations watched as the "sacred season of plenty" melted away. After a little more than a century of commercial fishing, many contend there are only remnants of the Pacific Herring population that once existed. One thing is certain, though: as of 2021 there is only one commercial herring fishery left in BC – the one in the Strait of Georgia.

The Strait of Georgia herring run is the largest of five major herring populations, the others being on the Central Coast, Prince Rupert, Haida Gwaii, and the west coast of Vancouver Island. Even with the decrease in historical abundance, the draw of other species to this banquet is still impressive.

BLACK BEARS FORAGE IN THE INTERTIDAL ON RICH HERRING ROE. PHOTO ERIC GROSS.

Don't be surprised if your own emotions run high when seeing Surf Scoters diving for herring, or a Black Bear foraging for eggs on the beach. Wildlife viewers should know a few things in order to improve their chances of having a successful viewing experience:

→ All of the beaches between Parksville and Cape Lazo in Comox are fairly reliable locations to witness this spectacle, but timing is critical. Although early March is generally a good estimate for the timing of the spawn, it can definitely vary. Herring are the decision makers here, but DFO provides Fishery Notices as a resource for those who rely on herring, and these notices contain information on the latest spawning activity, as well as on the locations of bird and mammal congregations that are associated with spawning activity. The notices begin to be provided in late February. Consider reviewing these to adapt your timing.

→ If you choose to view the spawn from a boat, consider reducing the noise made when viewing. Herring are an acoustic species and research shows boat noise is a significant disturbance for them.

→ Other species also plan their season around herring, so whether you are on the beach or the water, follow the ethical viewing guidelines outlined in the introductory chapter of this book to allow species to feed without interruption from humans.

→ The 13 Moons of the W̱SÁNEĆ website suggests two questions you may wish to think about when viewing the herring spawn: What signals does nature give us to start or end things? In what ways do you and your family prepare for spring?

Pacific Herring can live up to 12 years, but I'm told they are hard to find at even half that age because of all the species that prey on them. They feed on zooplankton (primarily krill), which vary with changing ocean conditions, such as temperature or currents. As the quintessential *forage fish,* herring as prey begin when females lay as many as 20,000 eggs! Only about one in 10,000 eggs survive to adulthood; the remainder become food for other species. Entire schools of herring can reach several kilometres in length. The Georgia Strait herring begin to migrate in late autumn en masse from the deeper ocean waters off the west coast of Vancouver Island to breed, arriving in early spring to the area around Hornby Island. In fact, they complete this migration to spawn multiple times throughout their lives, and when they do, they can completely cover nearshore bays and intertidal areas in a chalky white substance called "milt," which is the sperm-containing liquid expelled by a male fish. In these sperm-laden opaque waters, females lay sticky eggs on intertidal and nearshore vegetation, such as eelgrass and kelp. One square metre of eelgrass can be covered with as many as six million eggs![6] It's on the herring's schedule that *other species time their migrations,* and it is here in early spring that gulls, scoters, grebes, loons, eagles, and so many more birds congregate. This is truly one of nature's great spectacles! But the spawn is short-lived, lasting only days for the herring and a few weeks for the others they feed, and the timing is not entirely predictable. Like salmon, Pacific Herring go back to their natal areas to spawn, and herring eggs are vulnerable to water pollution. The result is that the overall resilience of the species is reduced.

Herring play a critical role in BC's coastal food webs and ecosystems. They are a dietary staple for many other fish, including cod and halibut (40 per cent of their diet), as well as flounders and hake. Trends in herring

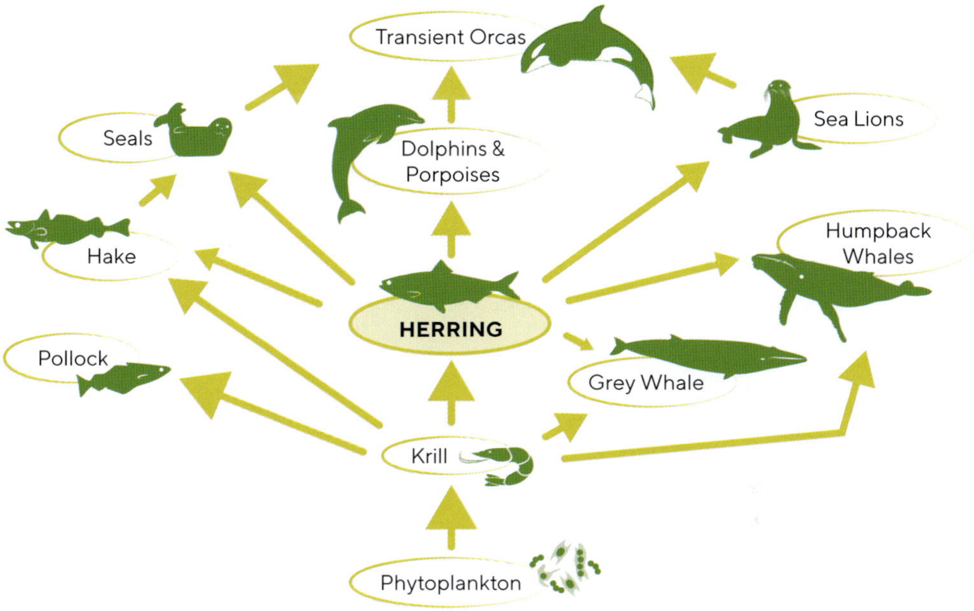

THE IMPORTANCE OF HERRING IN THE DIETS OF THEIR PREDATORS IS WELL DOCUMENTED. THIS IS A SIMPLIFIED DIAGRAM SHOWING THE MAJOR INTERACTIONS INVOLVING HERRING. SOURCE: FISHERIES AND OCEANS CANADA.

biomass have implications for their predators, some of whom have increasing populations while others are decreasing. Consider marine mammals that are significant predators: California Sea Lions, Harbour Seals, and Humpback Whales are all **high trophic level** species (species such as marine mammals that eat high on the food chain), and each is a success story of increasing numbers over the last several decades. But the Harbour Seal and California Sea Lion populations are increasing, while salmon numbers are decreasing. And these marine mammals eat salmon. This simplistic view referred to earlier would claim that if we cull the seal and sea lion numbers then the number of salmon available for ourselves will increase. Add in that herring make up 62 per cent of a Chinook Salmon's diet, and while herring have decreased, so too have Chinook populations. It doesn't add up. In my research I read more than once, "we can't save salmon if we don't save herring, and, if we don't save herring, we can't save the Southern Resident Killer Whales." (Refer to the chapter on Killer Whales, also known as Orcas, off Saturna Island to better understand the importance of Chinook Salmon for resident Orcas to flourish.) But if you implant any idea from this

A SEA LION AND GULL FRENZY INDICATES THAT THERE ARE LIKELY HERRING SCHOOLING BELOW THE SURFACE. ***IMAGINE THE SOUNDSCAPE THAT GOES ALONG WITH THIS SCENE.*** PHOTO TERRY CARR.

chapter into your mind, remember Scott Wallace's notion that "ecosystems are so much more complex than they appear." Sand Lance, Northern Anchovies, and Pacific Coast Squid are just a few species that are present in the same food web as herring and Chinook, and the abundance of each, or lack thereof, is a lever that affects the wider system. Bottom line: It's complex!

If you look at the BC coastal marine food web in this chapter, you'll begin to understand the complexities that exist, and the diagram doesn't even begin to address humans or their unseen impacts that affect the food web, such as ocean warming and acidification due to climate change, extreme acoustic disturbance, loss of access or damage to natal areas, the impacts of salmon farms, and so many other existing variables. In the BC case, if we were to cull sea lions, for example, in order to have more herring for humans to harvest, we might be dismayed to find that some or all of the five other herring predators had beaten us to the "extra" herring!

> Pacific Herring interact with nearly every other species in Canada's Pacific waters as they develop from egg to spawning adult and move from their nearshore birthplace to open ocean feeding grounds. Their abundance is primarily influenced by oceanographic conditions. The complexity of

BRANT GEESE AND GULLS MAY BE SEEN FEEDING ON HIGH-CALORIE HERRING ROE AT THE SHORELINE. ***LIKE FINDING WALDO, CAN YOU SPOT THE AMERICAN WIDGEON DUCK?*** PHOTO ROGER MEYER.

these ecological interactions hampers our understanding and confirms the need for managing these amazing fish with caution. We are fortunate that every year Pacific Herring make a brief appearance into the human realm when they come to shore to spawn. It's a globally significant spectacle worth viewing.

—**SCOTT WALLACE,** marine ecosystem specialist at the David Suzuki Foundation

Engaging and Taking Action

Project Watershed is a citizen science collaboration with North Island College wherein student volunteers work to identify forage fish spawning on beaches in the Northern Salish Sea (Strait of Georgia). Volunteers sample beaches for the presence of eggs of various forage fish species to help in our understanding for the recovery of other marine species at risk.

Eelgrass is prime intertidal habitat for Pacific Herring to spawn on, but much of coastal BC's eelgrass has been lost, particularly in coastal residential and recreational areas. Consider volunteering for the SeaChange

Marine Conservation Society for the sake of herring, or of the Orcas. (Refer to the chapter on Orcas on Saturna Island.)

Minimizing Your Carbon Footprint

Our carbon footprint is directly connected with the health of the oceans. Taking action to reduce it is an acknowledgement that herring health is impacted by our decisions. Island Link Bus runs a scheduled bus from Nanaimo's Departure Bay to the Buckley Bay ferry terminal, which will take you to Denman and Hornby islands. Once on the Denman ferry, I suggest hitching a ride to Hornby by asking someone on the ferry (I'm told they almost never see someone stranded). If on Hornby in summer, you can get around on the Hornby Bus for free. Both islands are bike friendly if the weather is decent, and on Hornby Island it often is!

Directions and Accessibility

From Nanaimo follow Highway 19A toward the Buckley Bay ferry terminal, which provides ferry service to Denman and Hornby islands. Consider stopping at the beaches between Nanoose Bay and Comox: Qualicum National Wildlife Area, Rathtrevor Beach Provincial Park, French Creek, Qualicum Beach and Bay, Deep Bay, Fanny Bay, Union Bay, and Royston. Hornby Island's Tribune Bay, Galleon Beach, Fossil Beach, and Helliwell Provincial Park are a few options for nature viewing from both low and high elevation points.

Accessibility varies depending on the community or beach area visited, but those areas around Parksville and Qualicum Beach are especially suitable for wheelchair access. Tourism Vancouver Island is a good contact for more information.

Other Locations to view Pacific Herring

The Strait of Georgia has many locations to view herring, but the key is the timing in the early spring. Areas on the Sunshine Coast and the east side of Vancouver Island north of Nanaimo are best. Contacting people or businesses that recreate or work on the water in specific communities to ask them what they are noticing regarding the abundance of wildlife in early March would be useful before making the trip.

OCHRE STARS, WHICH INCLUDE BOTH ORANGE AND PURPLE INDIVIDUALS, CAN OFTEN BE FOUND CLINGING TO THE ROCKS AROUND STANLEY PARK DURING LOW TIDES. IT IS NOT YET KNOWN WHAT CAUSES THE DIFFERENCE IN COLOUR. LEATHER STARS CAN ALSO BE SEEN AT THE BOTTOM RIGHT OF THE PHOTO. PHOTO DEBORAH SIMPSON.

Other Area Interests

1. From late April until about the May long weekend, Vaux Swifts visit the chimney of the Courtney museum and the old school on Maryport in Cumberland. Often, volunteers even have lawn chairs set up to watch on that weekend! They come in at dusk and there can be thousands of them. This is a spectacle in its own right.

2. The Coming of Signs of Spring: The Winter Birds chapter in this book introduces the idea of *phenology* and provides a suggestion to begin a journal in early spring to record what is happening in the natural world around your home and community. Early March is not just a great time for viewing wildlife around the Pacific Herring spawn, it is also a fantastic time to start observing the changes that are beginning with plants, birds, mammals, and other organisms.

Additional Resources

1. **HerringFest** – This event is part of Conservancy Hornby Island programming, which supports local and regional conservation projects that benefit Hornby Island's land and marine environments.

2. **Herring Roe Fishery, Strait of Georgia 2020** is a YouTube video by Bob Turner that accompanies the content of this chapter well.

3. **Shifting Baselines** – This is a theory hypothesized by UBC's Daniel Pauly, a frequent speaker at HerringFest. The theory suggests that when we measure change in a system, we typically measure against previously known reference points (baselines), which themselves may represent significant changes from an even earlier state of the system. Consider a natural area that you know, and how it has changed over your lifetime. Then consider how your parents, grandparents, and great-grandparents (three generations) may have experienced the same natural area. Changes we can measure in each generation are limited by our lifetimes. To be sure, ecosystems will vary with or without humans, but sometimes the evidence collected over time points to human-caused change. Herring live most of their lives out of our purview in deep ocean ecosystems west of Vancouver Island, but for a short time they visit "our" shorelines by the billions, in order to reproduce. Authorities set what they *think* are sustainable catch limits, based on the best available knowledge but still acknowledge that their data is limited in scope. Although the science

we practise and our understandings of the natural world have improved significantly over generations, we also need to acknowledge the bigger changes that have happened over multi-generational time, and that there are complexities inherent in ecosystems that we have yet to learn.

4. **The Raincoast Conservation Project** – The *Raincoast Conservation Foundation* knows that Pacific Herring are a cornerstone of the marine food web and have sustained coastal First Nations communities for thousands of years. The organization is using science to map out the ecological relationships of herring, as they aim "to understand the ecological consequences of the herring spawn and conversely, what the loss of herring means to coastal ecosystems."[7]

5. ***Gambling with Nature: Seals and Salmon*** is a YouTube video, by the David Suzuki Foundation, that provides context for how complex nature is, and the variables involved.

CLIMATE CHANGE AND PACIFIC HERRING

Climate change is causing warming waters off the BC coast, while creating other impacts in the estuaries surrounding the natal herring streams. As mentioned in this chapter, there are so many complex variables when it comes to herring that a changing climate adds yet another layer, a very significant one, into this already complex system.

THERE ARE MANY SIZES OF BOATS IN TOFINO TO CHOOSE FROM FOR YOUR GUIDED WHALE WATCHING EXCURSION. SOME, LIKE ZODIACS, OFFER A LITTLE MORE ADVENTURE IN THE EXPERIENCE! PHOTO REMOTE PASSAGES MARINE EXCURSIONS.

Grey Whales & Marine Mammals

in Clayoquot Sound & Pacific Rim National Park, Vancouver Island's West Coast

AT A GLANCE

WHAT: Grey Whales (*Eschrichtius robustus*), Humpback Whales (*Megaptera novaeangliae*), Sea Otters (*Enhydra lutris*). Look also for Sea lions, Seabirds

WHERE: Tofino, Long Beach, Clayoquot Sound, Pacific Rim National Park

WHEN: March/April–Summer

TRADITIONAL TERRITORY: Yuułuʔiłʔatḥ (Ucluelet), nuučaańuułʔatḥ nisḿa (Nuu-chah-nulth), ƛaʔuukʷiʔatḥ (Tla-o-qui-aht), Ahousaht

ACCESS: Easy. Requires time on a boat.

Roughly 20,000 Grey Whales pass through what's known as the eastern North Pacific waters, and right by Tofino, BC, every spring on their 16,000–22,500 km round trip migrations (the exact distances depend on their starting and ending locations). One of the best times to view them is during the Pacific Rim Whale Festival, which takes place in March, when migrating whales arrive in the area. These migrants include about 200 "summer resident" Grey Whales, a subset of a larger population that migrates from Mexico, but this group typically doesn't continue their migration much farther north than BC, staying in the area until they leave for the south in October. In this area, you also have the opportunity to view Humpback Whales (April to October), Minke Whales (May to October), and Orcas (commonly seen, but the timing is less predictable), plus porpoises, sea lions, and otters, all of which call these waters home.

I had the pleasure of interviewing Jim Darling, a west coast whale researcher with more than four decades of experience studying the whales that frequent these waters. Jim considers Grey Whales an integral part of his community, and he stresses that many of the same individuals return to the bays and sounds of the west coast of Vancouver Island every year. They know every cove, sandbar, and rip current and many are well known to the local (human) community. Some are even given names, like Saddle and Blackjack, a pair of males that have been known to swim side by side on many occasions, for over 40 years They are like old friends.

To view the whales, there are more than a dozen local whale watching businesses (each with approximately 3–5 boats) that cruise with talented naturalists on board, introducing visitors to these whales. Over the past 40 years, Jim has watched Tofino's whale watching tourism industry grow to maturity. He has observed the operators' high standards and dedication and believes in the positive effects their environmental education messaging has on their clients.

Jim says all indications are that the Grey Whales have become habituated to respectful whale watching, as there has been no overt change in behaviour or habitat use over time. However, he cautions that we must always be aware of human activity levels around the whales and monitor for avoidance or other unusual behaviour.

The operators who are on the water daily are perhaps the best monitors. Some even incorporate a "research fee" into their ticket prices, which provides capital to pass along to local scientists in order to support whale research in the area, perhaps as a recognition that much of the knowledge they are able to convey to their clients was discovered or confirmed by marine scientists working in the area. This money helps to support the continuation of important research into whales and other marine mammals in the local waters.

So, while it may seem like a contradiction that in another chapter in this book I suggest only shore-based viewing for Orcas, in this section I encourage the reader to join a local boat tour operator in Tofino or Ucluelet. A key difference is that to date current research hasn't found that the present levels and type of boat traffic, with their associated noise, have had a significant impact on the Grey Whales that pass by the west coast of Vancouver Island. In addition, in this area there's not the concern of large container ships passing through a narrow shipping lane, as is the case in the Salish Sea in Juan de Fuca, Haro and Georgia straits.

The whale watching season in Tofino begins in March with the arrival of males and newly pregnant females, followed by mothers with newborn calves in late April, through until June, as they return from the nursing lagoons of

A GREY WHALE "SPYHOP" ALLOWS THE ANIMAL A VIEW ABOVE WATER, A BEHAVIOUR PROBABLY AKIN TO US POKING OUR HEADS UNDERWATER TO TAKE A QUICK LOOK.
PHOTO TYSON CROSS / REMOTE PASSAGES MARINE EXCURSIONS, TOFINO, BC.

Mexico's Baja Peninsula. Calves stay close to their mothers as a protection from Orcas, their main predator, continuing to nurse until they wean at about 9 months. They don't reach puberty for 6–12 years, they keep growing until about age 40, and some may live for over 80 years![8] As adults the majority of these migrating 30-tonne *"baleen" whales* (whales that filter small marine organisms through baleen mouth plates) move north to their nutrient-rich summer feeding grounds along BC's coastal waters and as far as the Bering, Chukchi, and Beaufort seas, between or north of Russia and Alaska.

WHALE TAILS OFTEN DISPLAY UNIQUE MARKINGS OR PATTERNS THAT HUMAN VIEWERS CAN USE TO IDENTIFY SPECIFIC INDIVIDUALS. *LOOK FOR PRINTABLE ID GUIDES AT WILDWHALES.ORG.* PHOTO REMOTE PASSAGES MARINE EXCURSIONS.

Getting to know more about the whales before your visit will be an advantage once you find yourself crashing through Pacific waves on a wildlife watching boat excursion. To begin with, Grey Whales are medium to dark grey, barnacle encrusted, as long as 14 m (2 m longer than a school bus!), with vertebral bumps called "knuckles" (which take the place of a dorsal fin) on their lower backs. They forage unobtrusively in shallow waters (sometimes they're so still on the surface they are known as "breathing rocks") to about 60 m depth, and have specific behaviours at the surface and when diving. These include taking two or three surface breaths and flipping up their tail, before disappearing underwater for approximately three to five minutes. (To identify individual whales, bring a copy of the BC Cetacean Sightings Network printable guide, available at wildwhales.org.) When Grey Whales dive, they might feed on planktonic and larval species in the water column, or they may descend to a muddy or sandy sea floor where they scoop up mouthfuls of sediment, sieving water past crustaceans, amphipods, larvae, and plankton, through fine-toothed comb **baleen**, and finally pushing the residue with their tongue, to filter out the water through their baleen, before swallowing. This can amount to a daily food intake of 400–1000 kg, helping the whales replenish their blubber layer, diminished from their activities the previous winter. Grey Whales feed on up to eight species of plankton, but they have a preference for two or three. One preference is "mysid," which are shrimp-like crustaceans. Mysids rely on a close interrelationship with kelp, and cold-water temperatures are critical for thriving kelp beds, meaning that a warming ocean from climate change is likely to have direct negative impacts on this critical plankton species that Grey Whales prefer.[9]

Other local threats to the whales include getting tangled in discarded seine, gillnet, and longline fishing gear; oil spills in their foraging areas and subsequent habitat deterioration of coastal marine ecosystems; and, outside of BC, and potentially on the migratory route of a whale, coastal exploration by the oil industry (and their associated seismic blasts), which may affect whale communication and can harm their hearing. Currently, however, the Grey Whales off Pacific Rim National Park are considered one of BC's good news recovery stories (Sea Otters, sea lions, seals, and Humpback Whales are others), and the Greys are generally considered a population of over 20,000 individuals – all descendants of the roughly 4,000 whales that remained after commercial whale hunting ended in 1937. However, the stability of the population may be in question due to a situation that has been declared an **unusual mortality event** – a striking increase in Grey Whale strandings (a total of 608 strandings between January 2019 and December 2022, including 25 in BC) that have occurred along the west coast of North America from Mexico through to Alaska.

HARBOUR SEALS ARE AMONG THE MOST ENCOUNTERED AND EASILY VIEWED MARINE MAMMALS. *IT'S NOT UNCOMMON TO HAVE SEALS WATCHING YOU, TOO, WHILE YOU'RE OUT ON THE WATER.* PHOTO ROB ALEXANDER.

The Nuu-chah-nulth First Nations have a long and complex history around hunting Grey Whales, although their "catch" amounted to only a small number of whales per season – nothing compared to the half-million whales killed in the North Pacific by the whaling industry in the 20th century alone![10] Nuu-chah-nulth culture coalesced around the major food sources of whales, salmon, and berries, about 7000 BC. Nuu-chah-nulth stories and archeological evidence show whales provided not just food, but sinew, oil, and bone to craft technologies critical to their survival.

I was fortunate to take a boat trip with Moses Martin, elected Chief of the Tla-o-qui-aht First Nation. Moses helped me understand the spiritual development of his ancestors in hunting whales. The hunters had rituals that included fasting, bathing, abstaining from sex, and scouring themselves with shells, all in a process that may have taken up to eight months. Only the worthiest suitor would be "chosen by the whale" to allow the hunter to kill it. From elaborate canoes about the size of the whales themselves, and dug out from a single cedar tree, the hunting party used harpoons with bark rope tied to 20–30 inflated seal skins. Once the harpoon impacted the whale, the seal skins acted as floats, preventing the whale from diving. After hours, the whales would exhaust themselves to a point that allowed the brave hunter to jump into the water and sew the whale's mouth shut, preventing it from taking on water and thus sinking to the bottom. Jim Darling told me of a visit he had to Echachis Island, across from Tofino's Tonquin Beach, where he could see evidence of Humpback and Grey whaling with some bones that ultimately carbon dated to be as old as 1500 years.

Though there is no Grey Whale hunt by First Nations in Canada today, these animals are still at risk, as they require no passport and frequently travel between the BC coast and Neah Bay in Washington state. In the USA, the Makah Tribe (also Nuu-chah-nulth) has the right to traditionally hunt for a specific number of whales, so feasibly, Jim told me, the whales could be protected as an endangered species in the morning and yet be hunted in nearby (USA) waters in the afternoon.[11] A great read on this topic is Judith Lavoie's article in *Hakai Magazine*, "Protected on one side of the border, hunted on the other." It is a complex story that involves courts, environmental advocacy groups, politics, and endangered-species listing status of the whales in multiple jurisdictions. There is valuable learning in this process of re-establishing lost traditions and revitalizing culture, while honouring the science of what we know about whales today. Regarding this, it's worth articulating a concept called *Two-eyed Seeing*. This is a process that embodies both Indigenous and Western world views, acknowledging that no single perspective is better. The idea of two eyes "refers to learning to see from one eye with the strengths of Indigenous knowledges and ways of

A HUMPBACK WHALE AND HER CALF ARE EXCITING EVIDENCE OF A POPULATION RETURNED FROM NEAR DECIMATION. ***AS WELL AS THE BEND IN THE BACK, PAY ATTENTION TO DORSAL FINS TO HELP YOU DIFFERENTIATE BETWEEN HUMPBACKS AND GREY WHALES.*** PHOTO SABINE DECAMP.

knowing, and from the other eye with the strengths of Western knowledges and ways of knowing...and learning to use both these eyes together, for the benefit of all."[12]

The UNESCO Clayoquot Biosphere Reserve received its designation in January 2000. By spending time in the reserve, you can witness both the terrestrial and aquatic biological wealth of this region. That wealth is born out of long-standing respectful relationships and interconnections of people with the land and its species, starting with Nuu-chah-nulth First Nations such as the Tla-o-qui-aht, Hesquiaht, Toquaht, Ahousaht, and Ucluelet, who have lived on this land for millennia. An interpretive sign on the walking trail at Tofino's Tonquin Park asks you to "Naʔaataḥ" or "listen carefully with your whole being." Listen "to the ones who are now quietly speaking... listen and observe the plants, animals, forest, ocean, sky and heavens – that reality which is the source of life. This is how to learn, and how to begin to know and understand what it is that needs to be done." You may also come across "Cawakniš," the idea that "we are one, interconnected," or "Hishuk ish ts'awalk" – "everything is one," a concept that reflects "the interconnectedness of all things: species, ecosystems, and humans." This is also the guiding principle of the Clayoquot Sound Biosphere Reserve and a central tenet of the Nuu-chah-nulth First Nations.

Moses Martin, the Tla-o-qui-aht Elder and Chief, told me that the way we approach the world always comes back to one thing: respect. "It doesn't matter what you do in life, you always do it with respect." He also impressed upon me how important the revitalization of their language is, and that theirs is a dialect created over millennia and focused precisely on the place where the Tla-o-qui-aht people live – it is a "place-based" language. Words in the language are descriptive of the role each thing or being plays, or of how it came to be. They contain both history and perspective. Watch and listen for "c'ixwatin" (Bald Eagle), "qu?ušinmit" (Raven), and "waa?it" (Pacific Tree Frog). Or put your hands on "Sučas" (tree), whose name means "land-holder," a name that contains the teaching of the importance of trees, because without trees, the ground would slide away! In the previous sentences you will notice characters that are not part of our 26-letter alphabet. They are from the International Phonetic Alphabet (IPA), and the strange characters each represent a particular sound. A worthy challenge to deepen your connection to this place, even as a visitor, would be to spend some time learning the Nuu-chah-nulth language symbols and sounds, so you would know how to attempt to say the local words. If you're up to it, start at Firstvoices.com. Another challenge might be to time your visit to the west coast to take part in an educational program. Information about ecology, wildlife, skill development, and Indigenous courses are accessible in one place at West Coast N.E.S.T.. The courses are offered by a collection of organizations, cultures, and communities, convened through the Clayoquot Biosphere Trust, that

IDENTIFICATION OF A WHALE IS OFTEN BASED ON THE SHAPE OF ITS DORSAL FIN. CAN YOU IDENTIFY THIS WHALE? PHOTO ERIC GROSS.

are committed to sharing knowledge and building a stronger understanding of one of the last great natural environments on the planet.[13] Other notable learning opportunities include the interpretive programs at Pacific Rim National Park Reserve, a guided walk on Ucluelet's Wild Pacific Interpretive Trail, and a visit to the Ucluelet Aquarium, a non-profit that offers incredible value in such a small space!

When you visit the west coast, it is wise to widen your scope and consider allowing time for extended beach walks, for standing serenely on a boardwalk under an old-growth canopy, and for rambling along the rugged Ucluelet coastline, attempting to determine the answer to a quintessential coastal question, "Is the tide coming in or going out?" I would be remiss if I led those "wider-scoped" visitors to the west coast in the spring for Grey Whale viewing but didn't highlight the explosion of other wildlife in Clayoquot Sound at that time of year. If you're watching whales from a boat, then it's likely you'll see a range of other west coast mammals, including other whales (Humpback or Minke), Orcas, porpoises, dolphins, California or Steller Sea Lions, Harbour Seals, and Sea Otters; and on the shoreline, deer, river otters, and Black Bear (especially when the tide is out and they're feeding in the intertidal zone). Birds will also be abundant early in the year, as spring is the season of migration, and common sightings will likely include Tufted Puffin, Bald Eagle, Oystercatcher, Common Murres and Marbled Murrelets, shorebirds, and much more. If you are a basic birder and wish to bring your birding skills to the next level, consider joining a "pelagic" (ocean-going) birdwatching tour with one of the local Tofino wildlife viewing outfitters. For birders, the peak season of migration is slightly later in the calendar year than it is for Grey Whales, so you may wish to consider a trip between mid-April and mid-May. Early May also happens to coincide with the Tofino Shorebird Festival, run by the Raincoast Education Society. The west coast is a place to arrive prepared, so binoculars, field guides, and some pre-trip research will go a long way in readying you to observe wildlife in one of the most productive wildlife viewing sites in Canada!

> The summer Grey Whales are here only because of a handful of tiny prey species whose collective abundance make it worth their while. If the beds of bottom dwelling amphipods, or the swarms of planktonic mysids or crab larva disappear, so do the whales. Period. Our knowledge of these species is very limited – but we know they are susceptible to changes in their environment. The odds that an increase in sea temperature will have a significant impact on them are very high.
>
> —**JIM DARLING,** researcher, Pacific Wildlife Foundation

Engaging and Taking Action

Are you familiar with the Surfrider Foundation? The foundation is North American in scope but has a Vancouver Island chapter that offers monthly beach cleanups. Check out their website for scheduled cleanups or to volunteer to lead one. If none of the events fit your schedule, consider their actionable hashtag #5minutebeachcleanup. Beach cleans are especially important for wildlife that use beach and rocky shorelines (such as breeding birds, seals, and sea lions). Download the Marine Debris Tracker mobile app and use the Surfrider Canada list to upload debris recovered and removed from a coastal area. If your interest is more global, review the non-profit called the Ocean Cleanup, with a grand vision to remove 90 per cent of floating ocean plastic by the year 2040!

Pay an online visit to the non-profit Ucluelet Aquarium to see what programming they have scheduled. The aquarium pairs with the Strawberry Isle Marine Research Society to conduct monthly sea star surveys, and they

CONSIDER DOING YOUR OWN BEACH CLEANUP, AND IF YOU LIKE, *SHARE YOUR EXPERIENCE WIDELY TO ENCOURAGE OTHERS TO DO THE SAME.*

VISIT THE INCREDIBLY KNOWLEDGEABLE VOLUNTEERS AT THE UCLUELET AQUARIUM. ***EVERY YEAR THE AQUARIUM RELEASES THEIR INTERTIDAL SPECIES BACK TO THE WILD, AND IF YOU'RE THERE AT THE RIGHT TIME, YOU MIGHT BE ABLE TO PARTICIPATE IN THIS.***
PHOTO GINGER NAVETTE, "UCLUELET AQUARIUM TIDAL POOL WALKWAY," WIKIMEDIA COMMONS, CC BY-SA 4.0.

are open to the public. The purpose is to help record the recovery from the recent sea star wasting epidemic in coastal BC, including in Clayoquot Sound.

Review the resources discussed in the Orca / Killer Whale chapter of this book, such as the Whale Trail, the BC Cetacean Sightings Network and their whale report app at wildwhales.org, and Happywhale as a place to record whale sightings. These are all conservation initiatives that include the public, as citizen scientists, in their missions.

Minimizing Your Carbon Footprint

If you're going to travel by fossil-fuelled car, as always be sure to carpool if possible. More advanced cyclists could bike from Nanaimo to Tofino, a distance slightly over 200 km, or you could use the Tofino Bus to take you to the west coast. Whatever combination you choose, minimizing having yet another vehicle in that peninsula corridor would benefit both the environment and ensure you wouldn't have to worry about finding parking in Tofino on a summer Saturday! Once you've arrived in the Traditional Territories of the Yuułuʔiłʔatḥ (Ucluelet), ƛaʔuukʷiʔatḥ (Tla-o-qui-aht), and nuučaan̓uułʔatḥ nism̓a (Nuu-chah-nulth) First Nations, consider cycling the 42 km route between Ucluelet and Tofino, 25 km of which is on the new multi-use ʔapsčiik t̓ašii pathway (pronounced ups-cheek ta-shee, meaning "going in the right direction on the path"). If you don't have your own bike with you, T'iick'in Ebike Rentals (pronounced t-eets-kin, meaning "Thunderbird") provides rentals at the Ucluelet/Tofino Junction on Highway 4 and out of Ucluelet. The company is owned and operated by a local First Nation family and provides multiple styles of bikes to choose from. Finally, Tofino itself also has a small transit system that runs between Tofino and Ucluelet, with stops at various beaches, trail heads, campgrounds, and more.

Directions and Accessibility

Take Highway 1 to Horseshoe Bay and the BC Ferry to Departure Bay in Nanaimo (or Tsawwassen to Duke Point). From Nanaimo take Highway 19A to Highway 19 to Highway 4 (just past Parksville) all the way to Tofino.

Most whale watching companies in Tofino have accessibility options. Check Tourism Tofino for more information. Pacific Rim National Park accessibility options are listed on the Pacific Rim website.

Other Locations to View Grey Whales

Ucluelet also offers Grey Whale watching opportunities, as do operators in Port Renfrew (west of Sooke).

Other Area Interests

1. Forest walks to waterfalls at Little Qualicum Falls Provincial Park or Englishman River Falls Provincial Park.

2. The general funkiness (mostly shopping) at Coombs (Goats on the Roof is a popular tourist attraction in this area.).

3. You will be in awe at the extremely large trees in the old-growth forest loop trails at Cathedral Grove in MacMillan Provincial Park.

4. For those interested in archaeology, check out the petroglyphs (rock carvings) at Sproat Lake Provincial Park.

5. Though it is a substantially long gravel road, it is worth checking into the Bamfield Marine Sciences Centre and the public programs they offer.

Additional Resources

1. **Be a Better Beach Goer** – Prior to a beach visit, review the poster produced by the Ucluelet Aquarium.

2. **West Coast N.E.S.T.** – Helps people find courses and transformative learning experiences with over 30 community organizations, businesses, and individuals on the west coast of Vancouver Island. Courses come from a collective of organizations, cultures, and communities committed to sharing knowledge and building a stronger understanding of Clayoquot Sound.

3. **Ocean Wise Sightings Network** – has a good whale report phone app worth checking out.

4. **Surfrider Foundation** – offers scheduled beach cleanups in coastal British Columbia and you can find their latest events on their website.

5. **DFO Marine Mammal Incident Reporting Hotline** – For emergency marine mammal incidents you observe, such as distressed or deceased marine mammals.

GREY WHALES AND CLIMATE CHANGE

Grey Whales feed on up to eight species of plankton, but they have a preference for two or three. One preference is "mysid," which are shrimp-like crustaceans. Mysids rely on a close interrelationship with kelp, and cold-water temperatures are critical for thriving kelp beds, which means that a warming ocean from climate change is likely to have direct negative impacts on this critical plankton species that Grey Whales prefer.

THE ROTARY INTERPRETIVE CENTRE AT THE GREAT BLUE HERON NATURE RESERVE IS STAFFED BY VOLUNTEERS WHO ARE HAPPY TO HELP YOU PLAN YOUR DAY AT THE RESERVE.

Pacific Great Blue Herons

at the Great Blue Heron Nature Reserve in Chilliwack

AT A GLANCE

WHAT: Pacific Great Blue Herons (*Ardea herodias fannini*)
WHERE: Great Blue Heron Nature Reserve in Chilliwack
WHEN: Mid-March to mid-June
TRADITIONAL TERRITORY: S'ólh Téméxw (Stó:lō), Coast Salish, Nuxwsa'7aq (Nooksack), Kwantlen
ACCESS: Easy. Highway and backroads driving and easy walking.

With a primordial call that harkens back to the Jurassic, a wingspan that casts a shadow that one could easily be enveloped by, and a focus of concentration that a Zen master would admire, the Pacific Great Blue Heron is truly a magnificent coastal BC bird. The 325-acre Great Blue Heron Nature Reserve in Chilliwack is known for its significant breeding colony ("heronry") of nesting herons. To begin nest building by February, this non-migratory heron subspecies, which range from 115 to 138 cm (or an average of about 4 ft. tall), begin flights into the reserve's mature trees, particularly its Cottonwoods, before the trees leaf out. With no leaves through March and early April, it continues to be easy to view the herons for several weeks, until they lay their eggs. (Once trees leaf out, nest viewing from the dike is more difficult.) Babies with voracious appetites arrive mid-May, by then under cover of leaves, and their recognizable squawking ensures numerous parental trips to gather food.[14] Watching parents approach a tree and navigate their 1.8 m (6 ft.) wingspan between the branches to delicately place their gangly legs onto a nest among protesting young is a sight to behold. The nest-fly-feed-repeat process continues through August when the last of the young depart the nest.

PACIFIC GREAT BLUE HERON CHILLIWACK

THIS CALENDAR FROM THE RESERVE PROVIDES A GENERAL UNDERSTANDING OF THE KEY DATES IN A GREAT BLUE HERON'S YEAR. THE RESERVE IS OPEN ALL YEAR TO WATCH THE BIRDS IN ALL THEIR ACTION! CREDIT: SOUTH COAST CONSERVATION PROGRAM.

The Great Blue Heron Nature Reserve has a network of over 10 km of trails with boardwalks, a lookout tower, viewing blinds, access to the Vedder North Dyke Trail, a salmon viewing platform, and pond viewing platforms, as well as the volunteer-run Rotary Interpretive Centre (free admission) with educational displays and gift shop. Scan for the nests near the eagle's nest in the southwest part of the reserve at the junction between the heron loop and the Vedder Rotary Trail. The Pacific Great Blue Heron is a bird that, with binoculars or a spotting scope, you can look at in impressive detail due to its size and tendency to be still, especially while hunting. This allows a wildlife observer to see the eye's yellow iris, the sharpness of its long, thick bill, the arcs and cambers in its blue-grey feathers, and, if it's a mature adult, its distinctive black plume extending from above its eye to the back of its head. Just try to see that much detail the next time you're watching warblers!

"An ecosystem very seldom goes back to what it once was," Fernando Selles, a retired soil scientist and a long-time volunteer at the reserve, told me when we were discussing the reserve's history. This landscape was heavily disturbed as a water-based military training facility in its former identity as Canadian Forces Base Chilliwack. The current colony arrived in about 1960 after a previous site the herons nested at was logged. Eventually this site will go back to wetland, but not the wetland of old, because now it's surrounded by farms, dikes, roads, and suburbs.[15] All large heron colonies in Metro Vancouver and the Fraser Valley have declined over the past 20 years due to an increase in a key predator, Bald Eagles, and the

BRING BINOCULARS, AS THE NESTS AND HERONS ARE HIGH IN THE TREE CANOPY; SHOWN HERE IN LATE APRIL, THE TREES HAVE NOT YET LEAFED OUT. *AMAZE YOUR FRIENDS BY DISTINGUISHING A MALE FROM A FEMALE: THE MALE BRINGS THE STICKS TO THE NEST AND HIS MATE WEAVES IT IN.*

ever-increasing human influences.[16] However, Ross Vennesland, a species at risk biologist with Environment Canada, believes the herons might be achieving an equilibrium with the eagles, as heron declines appear to have levelled off. The larger colonies of the past are now smaller but more stable colonies, perhaps an adaptation to ecosystem changes. In coastal BC there are about 4,000–5,000 breeding-age *fannini* subspecies of heron.[17] This

THE GREAT BLUE HERON NEST NUMBERS VARY YEAR TO YEAR AND CURRENTLY NUMBER ABOUT 100. *WITH AS MANY AS FOUR(!) CHICKS PER NEST, HOW MANY CAN YOU COUNT DURING YOUR VISIT?*

reserve has monitored its own numbers closely since 1997 and counted a high of 206 nests in 2004, dropping to 120 in 2013, and now 90–100 as of 2024. Although nest numbers have declined, it is believed the colony is now stable, and it's important to note that community support has never been greater. The Great Blue Heron Nature Reserve is a great community story. According to executive director of the reserve, Camille Coray, most residents of Chilliwack are supportive and believe in ensuring the health of this species. The reserve has 91 volunteers who monitor, count, welcome visitors, and educate children, as well as participate in work parties such as trail maintenance and invasive species removals.[18] If you can see yourself in such service-for-nature work, be sure to read the Taking Action section below.

Most wildlife species in this book come with an instruction manual of sorts. The instructions aren't really for the wildlife, but rather for us. Formally, it's called a **Species Recovery Management Plan**, and a blue-listed (**listed species** refer to ranks of a species' conservation status) species such as the Pacific Great Blue Heron has a thorough one. Administered by Environment Canada, the plan provides a recovery strategy to ensure the birds' future success. Broad strategies within it include habitat mapping, protection, and restoration. The strategies also incorporate threat mitigation, conservation, stewardship, and education. The City of Chilliwack recognizes the importance of the reserve for both the species and the public education opportunity it offers (17,000 visitors in 2020 alone), and therefore is a major funder. Personally, I like to know there are biologists, city employees, reserve staff, and volunteers with their fingers on the pulse of this species and its habitat. I hope young people looking for a career path see the value in the work being done in reserves such as this.

Please pay special attention to the trails that are closed during breeding season, as these go directly under the nesting trees and the reserve rightly prioritizes the birds over recreationalists. There is solid research showing that repeated human intrusions are linked to abandonment of colonies. Consider the parental requirement to ensure obtaining enough calories to feed up to four nestlings so they can fledge (leave the nest) by August. This is done by stealthily catching fish in the bordering Vedder River or capturing small mammals and amphibians on the adjacent farmlands. But when people visit parks, we expect to have the right to let our dogs run off-leash, don hip-waders and fish the river ourselves, canoe, kayak, or paddle board, and all of these activities potentially lead to interrupting the herons' hunting. The more disturbances the herons suffer the less likely they will be to be successful in getting those calories to the nest. That said, some do habituate to humans, but generally only where people are predictable. The dike

IT'S A SPECIAL THRILL TO WATCH A GREAT BLUE HERON PATIENTLY STALK AND THEN EAT A FISH AT LOST LAGOON IN STANLEY PARK. *THE EYES TELL THE STORY!* PHOTO MICHAEL SCHMIDT PHOTOGRAPHY.

BOTH PARENTS MAKE FREQUENT FLIGHTS TO AND FROM THE NEST TO FEED THEIR FAST-GROWING YOUNG. AS WITH MANY SPECIES OF BIRDS, *YOU CAN WATCH THIS ONLINE TOO, WITH VARIOUS LIVE CAMERAS, SUCH AS THE STANLEY PARK HERONCAM.* PHOTO TIM LUMLEY, "GREAT BLUE HERON FAMILY," FLICKR, CC BY-NC-ND 2.0.

that borders the reserve offers one observing location that appears to have limited or no impact on the herons.

After closely observing the physical characteristics of a heron, consider watching its behaviour. It may be preening, moving sticks in the nest, hunting in the adjacent field, or showing signs of disturbance. If disturbed, it begins watching the intrusion closely by standing up, making a cluck-cluck sound, and, if the disruption continues, it may eventually flush (leave a nest). Of course, it's not just people who can disturb the heron; the birds are also constantly on the lookout for predators, which leads to a very interesting story, that of "**Mafia Protection**." It goes like this. Bald Eagles prey on heron eggs and nestlings, which one would think would make herons avoid eagles at all costs. Yet research shows that when an eagle pair moves their nest, herons actually pick up and move their own nests closer to the eagles, usually to within 200 m of the new eagle nest! They did this most recently in February 2020. That's right, they moved closer. What's up? It turns out Bald Eagles are highly territorial and defend the area around their nest vigorously. They too are raising young in the spring, and nearby heron nests are a relatively easy food source. The herons certainly pay a price by losing eggs and young, but the eagles, in defending their territory, keep other predators, including other eagles, hawks, ravens, and crows, at a distance. Therefore, only one eagle pair, raising one eagle family, has dining rights! The herons may begin with three or four young, but variables such as predation, inclement weather, human disturbance, or food availability often lead to an average of fewer than two juveniles fledging the nest.

Though the Cornell Lab's Allaboutbirds.org site displays a map showing that the Great Blue Heron can be found throughout most of Canada, the USA and Central America, it's worth noticing the *fannini* designation in its Latin name. This denotes a subspecies only found in and around coastal BC, and it's for that reason it is a blue-listed species in British Columbia and a COSEWIC (Committee on the Status of Endangered Wildlife in Canada) species of special concern in Canada. At the time of writing, we are still

RESEARCH SUGGESTS THAT HERONS MAY SET THEIR ROOKERY WITHIN THE GUARDED TERRITORY OF A BALD EAGLE PAIR, AS ONE SET OF MARAUDING EAGLES IS LESS RISKY THAN MANY. *WHEN VIEWING A HERON ROOKERY, WATCH FOR BALD EAGLES AND SEE IF YOU CAN FIND THEIR NEST AS WELL.* PHOTO ROB ALEXANDER.

WITH A 1.8 M WINGSPAN, GREAT BLUE HERONS ARE QUITE IMPRESSIVE WHEN FLYING OVERHEAD. PHOTO MICHAEL SCHMIDT PHOTOGRAPHY.

waiting and hoping for long-overdue Species at Risk Legislation in BC. Loss of biodiversity and the increasing impacts of climate change are the two biggest environmental issues of our time, and with no provincial legislation to protect listed species, such as the Pacific Great Blue Heron, we risk losing a focus on biodiversity as a priority. As you leave the reserve today consider scanning the surrounding landscape of the Fraser Valley. Watch for the competing priorities of farming, subdivisions being built near water, formal development of campgrounds along with informal local party sites, draining and infilling of wetlands, energy infrastructure placement, or the removal of riparian forest (the interface between land and a river or stream) for agriculture or homeowner view lines. In the Lower Fraser Basin from Vancouver to Hope, 85 per cent of bogs, marshes, swamps, and fens have been diked and converted to urban and agricultural use. Blue Herons are being pressured by human encroachment as they try to nest in areas where they can avoid human development while also remaining near suitable foraging habitat.[19]

Bald Eagles as heron predators are a known entity, but humans, they are the wild card! Humans often proceed with activities that are beneficial to themselves (a new dike here, a Vancouver port expansion there) and risky to wildlife (habitat loss here, a potential oil slick on the horizon there). As a powerful species, we don't take the idea of coexistence seriously. So where to go from here? The good news is that there are plans to improve the way we thoughtfully coexist with wildlife, and many dedicated people are already involved in carrying them out. And even better news: they are always looking for fresh recruits!

> I'm always amazed at the overall sensitivity of a species like this. How precarious it all is. Things can turn overnight: Recently, a wind and ice storm took out 50 nests alone! Blue Herons have been around for over 60 years, but we don't actually know if they will be here next year. One tree can have 20 nests, and if a beaver chooses that tree, that's 20 nests gone, just like that!
>
> —**CAMILLE CORAY,** executive director, Great Blue Heron Nature Reserve

Engaging and Taking Action

Though a wildlife observer may begin by watching the herons, it won't be long before they notice the vivid colours of the Wood Ducks or see the splash of a Beaver or river otter in the ponds. For those venturing indoors to the Rotary Interpretive Centre, there are historic photographs showing a greater density of trees (more trees = more heron nesting habitat) and a large, shallow, glacial lake where you are currently standing. Known as

BUSY RIVER OTTERS MAY BE FOUND EITHER ON LAND OR IN THE WATER BETWEEN THE DIKE AND THE GROVE OF TREES INSIDE THE RESERVE. ***WATCH AND LISTEN FOR THEM RUNNING ON LOGS OR SWIMMING IN THE WATER.*** PHOTO TERRY CARR.

BREEDING SEASON BRINGS OUT THE MOST FLAMBOYANT FEATHERS AND COLOURS IN MOST MALE BIRDS, AND THE GREAT BLUE HERON IS NO EXCEPTION. *USE BINOCULARS TO LOOK CLOSELY AT THEIR PLUMAGE, INCLUDING THE WISPY CHEST FEATHERS THAT THEY PREEN INTO THEIR OTHER FEATHERS.* PHOTO TIM CYR.

Sumas Lake, it was drained by the Vedder Canal to create fertile land for farming. These additional observations help broaden one's perspective, and this "bigger picture" helps move a casual observer along a spectrum toward becoming a reflective naturalist. You begin to see how precarious it all is – how just as COVID-19 in early 2020 changed lives almost overnight, so too can a nest-destroying windstorm or a change in policy around land use detrimentally affect the lives of wildlife in a short period of time.

If you, yourself, are a potential "recruit," or if you know other young, middle-aged, or older adults interested in gaining a wider perspective and seeing the overall sensitivity of this species, then be aware that there exist broad strategies for you to engage in. These include projects on habitat conservation, restoration, research, mapping, and monitoring, as well as education programs to help others learn, understand, and ultimately place greater value on these ecosystems. Just open the door to the Rotary Interpretive Centre and begin with your first question: "Excuse me, how can I help?"

Minimizing Your Carbon Footprint

Put your bike on one of the two bike rack spaces on the Translink R555 Port Mann Express from Lougheed Station to Carvolth Exchange at 202nd Street and Highway 1 in Langley. At the Carvolth Exchange transfer your bike to the BC Transit 66 FVX (Fraser Valley Express) bus. Take this bus to Vedder and Luckacuck. Then take Route 1 Vedder bus to Garrison Boulevard. Cycle 7 km west on the dike to the reserve. A second option is a scheduled charter bus called Ebus that can be found at myebus.ca.

Directions and Accessibility

Take Highway 1 to Exit 104, and go south on No. 3 Road. Go left on Boundary and right on Keith Wilson, and take a final right on Sumas Prairie Road. There you will find the Rotary Interpretive Centre, with washrooms, an outdoor classroom, and picnic tables and benches.

The reserve also has wheelchair-accessible buildings and trails. Search Chilliwackblueheron.com or email the Reserve and Rotary Interpretive Centre at herons@shawbiz.ca.

Other Locations to View Great Blue Herons

1. Stanley Park Tennis Courts at English Bay near the parks board office is a great centre of the city location to view herons.

2. Up to 700 herons (15 per cent of the entire Canadian population!) are at the Tsawwassen Bluffs and Jetty along the BC Ferries causeway. Scan the sand flats at low tide.

3. Look for a small heron colony on the north side of Port Moody's Shoreline Trail, near Old Orchard Park. The picture on the cover of this book is from that location.

4. McDonald Park on No. 3 Road in Abbotsford is likely the closest neighbouring colony to those at the Great Blue Heron Nature Reserve in Chilliwack.

WATCH FOR INTERACTIONS BETWEEN GREAT BLUE HERONS. AMONG THE MOST SPECIAL ASPECTS OF BIRDWATCHING IS *TAKING TIME TO OBSERVE BIRD BEHAVIOUR.*
PHOTO WALTER SUSSEL.

Other Area Interests

1. Check out Cheam Lake Wetlands Regional Park for birds, dragonflies, and a boardwalk wetland ecosystem.

2. Bridal Veil Falls Provincial Park is a 20-minute walk to a beautiful waterfall through a second growth forest.

Additional Resources

1. ***Nesting with the Devil*** is an 11 min. film on Vimeo about the connection between Bald Eagles and Herons.

2. ***The Hunt*** is a 3 min. film on Vimeo. It is the most tense, white-knuckle heron hunt you'll ever see. Excerpt from Mike McKinlay's film *The Perfect State*.

3. ***Before We Lost the Lake*** by Chad Reimer is a detailed cultural history and story of Sumas Lake.

4. **Adopt a Heron Nest** (symbolically, of course) through the Stanley Park Ecology Society (SPES), supporting their efforts to research, monitor, and protect Stanley Park's Great Blue Heron colony – a great gift idea for a child! Then watch them on the SPES's Heron cam.

5. **Cornell Lab's Allaboutbirds.org** – Earlier in the text of this chapter a website called All About Birds was highlighted. This is a great site to visit if you have a special interest in birds. On the site are maps, sounds, videos, and detailed information on every North American bird, including the Great Blue Heron.

6. **Cowichan Estuary Nature Centre Heron Cam** overlooks the largest heron rookery on Vancouver Island.

IT'S THE ORANGE FEATHERS JUST BEHIND THE WING ON THIS HUMMINGBIRD THAT HELP ONE IDENTIFY IT AS A RUFOUS HUMMINGBIRD. PHOTO JON MORAN.

Hummingbirds

in Princeton along the Kettle Valley Railway Trail

AT A GLANCE

WHAT: Rufous Hummingbirds (*Selasphorus rufus*), Anna's (*Calypte anna*), Black-chinned (*Archilochus alexandri*), Calliope (*Selasphorus calliope*)
WHERE: Princeton Kettle Valley Railway and the Trans Canada Trail
WHEN: First three weeks of May
TRADITIONAL TERRITORY: Syilx tmix^{w} (Okanagan), Nłeʔkepmx Tmíx^{w} (Nlaka'pamux)
ACCESS: Easy. Highway driving and easy walking.

Watching hummingbirds, as small as they are, is relatively easy and, because their territorial possessiveness is so obvious, their behaviour is fun to watch: feats of flight that include hovering (the only bird that truly does), flying backward, on their side or upside down, and, if you are especially fortunate, mating rituals of large loop flights complete with unique sounds and the pulling out of the loops at nine Gs of gravity![20] Close observation allows you to see the feathers in their "gorget" (the area below their bill, pronounced gor-jit) with its patterns and iridescent colours, which do not actually have colour pigment but rather reflect and refract sunlight to your eye, creating a brilliant hummingbird effect.[21]

Five species of hummingbirds are found in BC, and Princeton sustains four of them: Anna's, Rufous, Black-chinned, and Calliope. Although the town is often just a fuel break for many hummingbirds migrating north in early to mid-May, Princeton's food resources and geography are attractive to some for nesting; many also remain until late summer. Anna's Hummingbirds come from the coast by June for food resources, probably using the Similkameen River corridor, so you can also look for them at the Manning Park Resort feeders on their route. Rufous and Black-chinned Hummingbirds arrive on their migrations north from Mexico and the western United

THIS MALE CALLIOPE HUMMINGBIRD HAS A PURPLISH-RED STREAKED THROAT PATCH *GORGET*, MAKING IT UNIQUE AMONG HUMMINGBIRDS AND MORE EASILY IDENTIFIABLE. IT'S FEEDING ON THE NECTAR DEEP IN THE PETALS OF A COLUMBINE FLOWER.
PHOTO JON MORAN.

States. They arrive along with Calliope Hummingbirds, also from Mexico, a species that moves their way up to the higher mountain slopes. Males are drawn to flowers and feeders, while the females arrive looking not just for nectar also for high protein invertebrates like insects and spiders, to ensure the proper building materials for their eggs and young.

Princeton hosts one of BC's 37 **hummingbird banding stations**, and in 2019 over 2,000 hummingbirds were banded! Other stations are in the Okanagan, Kootenays, Vancouver Island, Fraser Valley, Vancouver, Omineca (Fort St. James), and the Cariboo in BC's north, that is, in virtually every north-south migration valley in the province. Banding is critical for providing details beyond what can be obtained through in-flight or feeder observation. Age, sex, quality of feathers, site fidelity, longevity, and – one thing that will be discussed in more detail below – toxin levels. Long-term detailed data collection is more critical now than ever in order to understand the bigger geographic picture. The big three in that picture are: (1) the birds'

THIS MALE ANNA'S HUMMINGBIRD IS PUTTING ON A DISPLAY FOR A FEMALE, RECOGNIZED BY THE SPREADING OUT OF ITS TAIL FEATHERS AND THE PUFFING UP OF ITS BRILLIANT PINK GORGET. PHOTO JON MORAN.

wintering grounds, (2) the nectar corridor of their migration route, and (3) their summer grounds. All four hummingbird species in Princeton cross many human boundaries and borders in this geographic big picture, and precarious situations can arise anywhere. Banding helps "provide information on the quality of the birds' breeding or wintering grounds, availability of food, and survival of both the young and adults."[22] Banding is regulated by the Canadian Wildlife Service (CWS). CWS protocols ensure bands don't interfere with a bird's ability to go about its daily activities by ensuring the band is no greater than 3 per cent of a bird's body weight, about the amount of a large meal.

Relative to its body length, the Rufous Hummingbird, at just the weight of a nickel, migrates farther than any bird in the world.[23] During its coastal migration of up to 4000 km between Mexico and Alaska, it follows the flowering of shrubs such as salmonberry northward through BC. In southwestern British Columbia, we are fortunate to have these jewels adorn our gardens between March and May, while they're nesting. Nests are built in the shape of a cup by the female, first weaving soft plant material, then moss, and all bound together by strands of spider webs. Her final touch is to add lichens for camouflage before she lays eggs and then, on her own, raises

TO HELP THEM UNDERSTAND HOW CHEMICALS MAY BE AFFECTING BIRDS, RESEARCHERS SOMETIMES COLLECT URINE SAMPLES SUCH AS IN THIS GENTLE PROCESS WITH A MALE RUFOUS HUMMINGBIRD. PHOTO ALISON MORAN.

HUMMINGBIRD NESTS ARE BUILT ON BRANCHES WITH MATERIALS SUCH AS LICHEN AND MOSS, AND BOUND WITH SPIDER WEBS. THESE TIGHT, TINY CUPS PROVIDE YOUNG HUMMINGBIRDS WITH NEEDED PROTECTION AS THEY GROW QUICKLY, MOSTLY ON A DIET OF INSECTS PROVIDED BY THEIR MOM. PHOTO SABINE DECAMP.

multiple young. In Princeton, however, their timing skews slightly down the calendar, and Rufous males can be seen from May right through August, while females and juveniles can still be seen at feeders until September. This timing varies considerably if you are in Metro Vancouver or southern Vancouver Island, but generally their order in departing from BC gardens is: adult males, then the first clutch, adult females, and finally, the fledglings of the second clutch. (A clutch, in birding terminology, is a group of eggs that a bird is raising, and some species raise more than one set of chicks in a season.) The slightly larger Anna's Hummingbird, Vancouver's official bird, resides year-round in southwestern BC, raising young as early as March.[24] If you already know this bird, no doubt you would say it is feisty, aggressive, or even downright hostile, but let's not anthropomorphize too hastily. Hummingbirds need unfettered access to nectar and invertebrates (insects and spiders) to keep up a rate of metabolism that is unmatched in the avian world. Consider its heart during flight (1,000 beats/min.), respiration (250 breaths/min.), wing stroke rate (70 beats/sec.), its long and precarious migration, and its ability to survive cold weather nights in **torpor** (a reduction in body temperature and metabolic rate). These exertions all require energy in the form of calories, and lots of them, and for an adult hummingbird, flower nectar is a major portion of that calorie source. Therefore, a hummingbird's aggressive behaviour can be interpreted as its way to protect its territory around that fuel resource.

If you have never visited a banding station, I strongly encourage you to do so, especially if you have a child interested in science. The close and personal interaction with a wild bird often has a lasting impact. For those interested in volunteering with birds, see the Taking Action section below. Overall, the number of Rufous Hummingbirds in BC is decreasing, while others, such as the Anna's and Black-chinned, are stable and expanding their ranges due to climate change. For a bigger picture on the status of birds in Canada, see the infographic-rich State of Canada's Birds 2019 report. But hummingbirds are the subject here, so a focus on the key issues surrounding them is necessary.

Dr. Alison Moran is a microbiologist and director of the Rocky Point Bird Observatory's (RPBO's) Hummingbird Project. Alison collects urine and feces samples from hummingbirds for environmental DNA ("eDNA") analysis, in order to understand the diversity of invertebrates and the associated toxins in their diet. In Greater Vancouver and the Fraser Valley, hummingbirds search out nectar and invertebrates in cultivated landscapes, including monoculture blueberry fields where agricultural pesticides (neuroactive insecticides called neonicotinoids, also known as "**neonics**") are often used to reduce insect pests. As a result, the amount of available insect protein, which the hummingbirds also seek, is reduced. This is compounded by additional evidence that these pesticides move into other species, such as bees, as well as into the nearby soil and water. Alison told me that a "baby hummingbird is grown on mostly soft and fatty smashed insects, while sugar (nectar) is just a bit of energy on the side." The insects have sublethal toxicity, but when hummingbirds eat thousands of insects, that toxicity accumulates. This was a completely new understanding for me, as I think of hummingbirds as nectar feeders primarily, but I learned how critical insects are for both juvenile development and for building strong feathers throughout their lives. The diversity of hummingbirds' invertebrate diet is immense, and when their foraging overlaps wetland and aquatic environments (which are particularly susceptible to toxicity concerns due to runoff), hummingbirds are at greater risk. Wetlands are critical habitats for many bird species as they are larval grounds for insects such as no-see-ums, midges, and moth, dagger, and dance flies, and these make up a significant portion of a hummingbird's diet.

Proper Feeding of Hummingbirds in southwestern BC provides nature-focused enjoyment, but it's also an ongoing responsibility that should not be taken lightly. If you put up a feeder, be very careful that your good intentions do more good than harm by following Alison's expert advice below:

- Bird feeders should be easy to clean; otherwise they can be a source of disease and contamination. Clean them every few days in the heat of summer and about once a week in the winter. Change the nectar before it's cloudy. Cloudy water means bacteria, which have likely been introduced by hummingbird tongues. Alison says, "If you wouldn't drink it, why expect them to?" My favourite feeder uses suction cups to stick to my office window, which is also great if you're interested in photographing the birds.

- Use ONLY white table sugar (NOT honey, brown sugar, icing sugar, juice, or commercial mixes with red dye). This is most similar to the pure form of sugar flowers would provide. *The recipe is on your hand in the form of four fingers (water) and one thumb (sugar).* Mix four parts *water* with one part *sugar* by boiling the water/sugar mixture and allowing it to cool before filling the feeder. Fill only as much as they will drink in a few days, keeping extra in the fridge to refill after cleaning.

- Feeding in winter and early spring certainly helps our local hummingbirds as it provides a nectar source when few flowers are available, and the feeders mimic an artificial "sapsucker well" in both winter and spring. (Sapsuckers are birds that make small holes in the bark of trees to draw nutrient-rich sap to the surface. When hummingbirds arrive in early spring, they access these wells, obtaining sucrose, amino acids, and minerals.)

- Please avoid feeding hummingbirds near a glass railing on a balcony as these can be dangerous for birds, tricking them into flying into the glass (they think they'll be flying *through* it) as they head for the feeder. Instead, if you place your feeder near a window, they will not be picking up speed as they approach, but rather slowing down to feed. The other place for feeders is protective greenery, such as shade trees or shrubs, but be careful there too because of potentially roaming cats.

When a hummingbird is feeding, watch it closely and you will see its long tongue flick out at up to 12 times a second. The tongue is split and unfurls from the bill, and as it retracts it acts like two pump cylinders that fill with nectar.

WINDOW FEEDERS PROVIDE MUCH ENJOYMENT TO THE VIEWER WHILE PURPOSEFULLY KEEPING BIRDS FROM FLYING INTO THE WINDOW: THE BIRDS SLOW DOWN TO LAND ON THE FEEDER, STILL A SHORT DISTANCE AWAY FROM THE GLASS. PHOTO CHRISTINE GAIO.

If these tiny aerial dancers seize your attention, then it's likely you will, like many others, start asking questions. As you do, you will learn that hummingbirds exist only in the "New World" (North and South Americas), and that they've been here for at least 10 million years! With the uplift of South America's Andes Mountains, new habitats were created, and hummingbirds evolved there, finding their own niches in these new environments, eventually diversifying into over 300 species! Later, the rise of Panama three million years ago created a land bridge that made it possible for the hummingbirds to access new habitat.[25] Further northern migration allowed access to the seasonal resources of North America. As flower and hummingbird species co-evolved, partnerships developed such that flower colours, nectar reserves, and sexual structures in certain plants made pollination by specific hummingbird species more likely. The birds evolved the massive brain (4.2 per cent of body weight – more than double that of humans!) necessary for their flying feats and enabling them to recall where the fuel resources (flowers and nectar) in their territories were more abundant.[26]

These are fascinatingly complex birds, and that complexity extends to **individuality** as well. Just as humans develop relationships within their communities and have a sense of place, so too do other beings, including birds. So, let's think about that. We tend to see species as a collection. The crows that sit on my fencepost are an assortment within their particular species. We think, "Today's crow can't be the same as the one sitting there yesterday, last month or last year, can it?" Is it possible that a particular crow visits that post on its territorial rounds daily, and in fact, knows that post far better than the homeowner ever has? Without obvious distinguishing features, yesterday's crow looks like today's, but it could be any crow, couldn't it? Wildlife researchers who band, tag, or watch a species faithfully notice individuality in species, and they would tell you that many individuals of a species are extremely faithful to their place in the world, just like you. This is known as "**site fidelity**," and hummingbirds have high site fidelity.[27] In fact, if you've had a feeder out for several years in a row, one of the hummingbirds at your feeder today is likely one of the same birds you fed last year, returning to the southwestern corner of BC, to your neighbourhood, to the northwest side of your house, to your particular feeder, and it probably even arrived in the same week it arrived last year! But this is just the beginning of its recall and spatial abilities. A hummingbird's memory and the calculations it makes are staggering. Research shows that Rufous Hummingbirds know the location of hundreds of flowers in their territory, as well as the bloom times and the fill rate of nectar (different flowers produce nectar at different rates). Effectively, their brain has a stopwatch of sorts for *each* flower, helping them gauge when to return to feed again. They manage *time and place* with a spatial acuity that ensures they don't waste critical metabolic energy searching for nectar that hasn't replenished, because they can't afford that. They remember not just the what, and the where, but the when.[28] Knowing what you do now, isn't it time to get to know an individual hummingbird that shares the space around your home?

> These blueberry fields that surround us may be green but they may as well be pavement for what they offer to the birds. For most birds, insects are a fundamental ingredient for raising young. Because of the pesticides used, there are very few insects and, therefore, few breeding hummingbirds.
>
> —**ALISON MORAN**, hummingbird scientist and researcher

SPIDER WEBS ARE CLEARLY BEING USED TO ANCHOR THIS NESTLING PAIR TO THE EVERGREEN. PHOTO MICHAEL SCHMIDT PHOTOGRAPHY.

A PAIR OF EGGS IN A HUMMINGBIRD NEST ARE EACH NOT MUCH BIGGER THAN A TIC TAC MINT. PHOTO MICHAEL SCHMIDT PHOTOGRAPHY.

Engaging and Taking Action

Do you have good eyesight? Precise and gentle hands? Attention to detail? Are you good at recording data? Do you love sharing the joy of nature and what we can learn from it with others? The Rocky Point Bird Observatory (RPBO) on southern Vancouver Island is an organization that has trained many volunteers and always requires committed assistants. Volunteering with RPBO can involve being a recorder, banding birds, mending mist nets, or interpreting and encouraging others to have an affinity for birds. RPBO runs in-person and online events as well, and always requires assistance around Zoom presentations, book and retail sales, and workshop organization. Giving time to support the science around birds is rewarding; it's hard not to make an emotional connection when you have a tiny bird in your hand.

Taking action to support hummingbirds can be as simple as setting up and supporting bird-friendly habitat. Hummingbirds are pollinators and provide an incredibly valuable ecosystem service. Support them by planting **hummingbird-friendly gardens**, free of pesticides and away from cats, and by creating water areas for invertebrates to thrive in your garden. If you choose to provide a year-round feeder for Anna's Hummingbirds, ensure it doesn't freeze through the winter and that it doesn't go empty. Placing the feeder out in the fall tells the birds that it is a reliable food source, and they will rely on it as such. Also consider participating in Birds Canada events such as Project Feederwatch (a critical research project to help winter bird populations), or the mid-February Great Backyard Bird Count, where you watch lists roll in from around the world as the data, including yours, shows as glowing lights on a real-time bird sightings map. Finally, begin to learn why birds are important for biodiversity, speak up for protection of their habitat and for wetland protection, and consider writing your Member of Parliament, requesting them to support the banning of neonicotinoids.

Minimizing Your Carbon Footprint

Princeton is 300 km and a four-hour drive from Vancouver. Mountain Man Mike's Bus Service is an alternative option to driving, taking just under four hours to travel between the Braid SkyTrain Station in New Westminster and Princeton twice a week. Check out their schedule online. The bus service accepts bikes for an extra fee. Consider cycling the area between Princeton and Tulameen (26 km and approximately 1.5 hours) or to other natural areas such as Swan Lake, China Ridge, or the Old Hedley Road next to the

Similkameen River. You wouldn't have to camp, however: Princeton is a very walkable town, with motels and camping opportunities. You could, using the bus schedule, arrive on a Thursday and return on a Saturday.

Directions and Accessibility

Take Highway 1 and Highway 3 to Princeton. To find hummingbirds around town, park at Burton Avenue and Barr Street, and walk the Kettle Valley Rail Trail, part of the Trans Canada Trail, through the tunnel. Here you will have to turn around and walk the KVR in the other direction. The gravel trail is mostly level, wheelchair accessible, and parallels both the Similkameen and Tulameen rivers, as well as the Cormack Marsh wetlands. In November 2021 this bridge was washed out by flooding. As of printing there is no set date on when the bridge will be replaced. The KVR trail towards Summerland is an alternate walk, as is Swan Lake. Bring binoculars, a bird field guide, and consider wearing a red hat (the colour often attracts hummingbirds), and you will potentially see Rufous, Anna's, Calliope, and Black-chinned hummingbirds along your walk in early to mid-May. The Vermilion Forks field naturalists have installed several

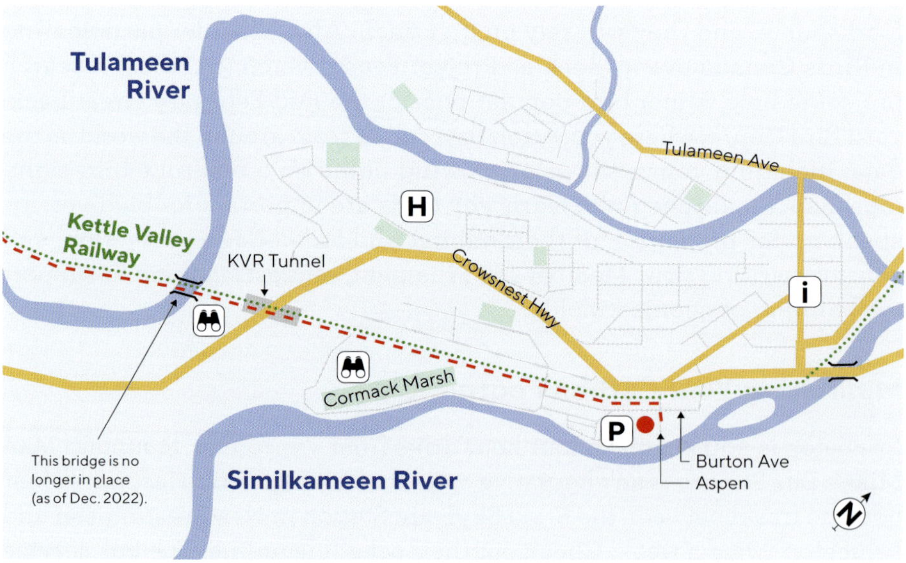

A GREAT LOCATION TO VIEW HUMMINGBIRDS IN PRINCETON IS ALONG THE SIMILKAMEEN AND TULAMEEN RIVERS ON THE KETTLE VALLEY RAILWAY TRAIL. SOURCE: PRINCETON TOURISM.

interpretive panels at the conservation areas, just off the Princeton-Summerland Road.

The trail is flat and mostly gravel and therefore reasonably accessible. Discover Princeton may have more information on accessibility options around town.

Other Locations to View Hummingbirds

1. Hummingbirds are common enough in southern BC that, once you know their sounds and behaviour, you should be able to spot them in your neighbourhood or local protected area with ease. From Princeton, another great location to view them is Otter Lake Provincial Park in Tulameen, approximately 30 km northwest of Princeton on Coalmont Road.

2. Though Anna's and Rufous hummingbirds have been the main focus of this chapter, southern BC is also home to breeding Calliope and Black-chinned hummingbirds. Both are commonly found in the Okanagan between Osoyoos and Penticton, while the Calliope is also found in mountainous areas such as Manning or Cathedral provincial parks. Calliope are the smallest birds in North America, and the smallest long-distance migrants in the world, travelling an 8000 km oval migration route each year while being just the weight of a Ping-Pong ball! Black-chinned are exceptionally widespread, found in deserts, mountain forests, and urban areas, but *most* of them are distributed south of the Canada–USA border.[29]

3. In southwestern BC, Vancouver's Queen Elizabeth Park, VanDusen, Stanley Park, and Victoria's Butchart Gardens all attract hummingbirds in the summer months. And remember, when hiking in southwestern BC alpine habitats, to be on the lookout for hummingbirds when alpine flowers are blooming.

Other Area Interests

1. Fossils – Check out the Vancouver Paleontology Society as they schedule trips around Princeton to uncover Eocene (34–56-million-year-old) fish fossils.[30]

2. Bromley Rock Provincial Park – A rock bluff along the Similkameen River that is great for summer swimming and camping.

3. Hedley's historic Mascot Mine is a window into some of this area's recent history as well.

4. Kettle Valley Railway to Coalmont is a lovely 18 km (one-way) trail on the KVR rail bed.

Additional Resources

IUCN Red List – The International Union for Conservation of Nature is a good additional resource. Search out the Rufous Hummingbird to further understand why its numbers are decreasing, and the actions and management policies that are being taken to help this species. The IUCN list comes up several times in this book (see Vancouver Island Marmot), and for someone interested in wildlife and the issues that surround it, it is a good resource to know.

CLIMATE CHANGE AND HUMMINGBIRDS

Hummingbirds migrate thousands of kilometres to reach flowering plants and insects to support their breeding, but when climate changes, these flowering or insect hatching times can be out of sync with their migration. Overall, Rufous Hummingbirds in BC are decreasing in number, while other species, such as the Anna's and Black-chinned hummingbirds, are stable in number and are even expanding their range due to climate change. It's useful to understand the bigger picture on the status of birds in Canada. To do this, see the infographic-rich State of Canada's Birds 2019 report.

HUMMINGBIRDS ARE GREAT SUBJECTS FOR NOVICE WILDLIFE PHOTOGRAPHERS TO BEGIN WITH, ALTHOUGH TO GET AN IMAGE AS SHARP AS THIS A VERY FAST SHUTTER SPEED IS REQUIRED. PHOTO JON MORAN.

***USE YOUR CAR AS A BLIND WHILE YOU TAKE YOUR BEAR PICTURE FROM INSIDE*, AS HAS BEEN DONE BY THE PHOTOGRAPHER WHO TOOK THIS IMAGE.**
PHOTO CHRISTINE GAIO.

Black Bears

in the Clearwater Corridor at Wells Gray Provincial Park

AT A GLANCE

WHAT: Black Bears (*Ursus americanus*)
WHERE: Wells Gray Provincial Park
WHEN: Mid-May to mid-June
TRADITIONAL TERRITORY: Secwepemcúl'ecw (Secwépemc), Dakeł Keyoh (ᑕᗸᒡ ᗕᘏᐦ), Dënéndeh, Ktunaxa ʔamakʔis
ACCESS: Easy. Clearwater Corridor: Spahats Falls to Clearwater Lake Campground.

Consider the last time, or the first time, you saw a bear. In that moment I'll bet you assessed it – how young it was, its size, its colour, and what it was doing. You didn't keep your experience to yourself either, but shared it with someone else. You crafted a story about the animal, the level of danger you were in, and how you came to encounter it. And today, regardless of how long ago that experience was, you remember it still, with all the details of how your body felt in that moment. That's the primal feel of meeting a bear, and in BC, if you live here long enough, you'll get to experience it and share your story. Of course, most of us feel different when we encounter a squirrel, a deer, or a heron, and it's likely because, when we encounter something as potentially dangerous as a bear, our brain brings forth a response (fear) that kept us alive in our evolutionary past.

The name *Ursus* (Latin for bear) may be familiar to you from the last time you looked up at the night sky and noticed the Big Dipper. A constellation map likely had the words Ursa Major or Great Bear, of which, humorously, the Big Dipper only constitutes the back end. Next time you look skyward in a dark location, challenge yourself to see all of the stars that make up the Great Bear. Back to Earth, it's worth noting that *Ursus americanus* (Latin for

IMMEDIATELY AFTER WE ASCERTAIN THAT IT'S A BLACK BEAR, THE NEXT THING WE TRY TO SURMISE IS THE BEAR'S AGE, BUT THAT'S NOT ALWAYS EASY. THIS IS THE WIDE-EYED, BIG-EARED FACE OF A YOUNG BLACK BEAR. PHOTO TIM LUMLEY, "AMERICAN BLACK BEAR," FLICKR, CC BY-NC-ND 2.0.

American Black Bear) is a species that resides just about everywhere in BC – on the coast, in the interior, up in the mountains, and down in the valleys. As these bears are omnivores (meaning they eat both plants and animals), vegetation makes up a large portion of their diet, though as with humans this can range wildly between individuals and between seasons. Nevertheless, all bears spend a considerable amount of time foraging for the foods that will get them through their **hibernation** from November to April. Bears don't eat, drink, or defecate during this time, but they are *not fully hibernating* either; instead, it is more like a deep sleep. Like other animals that hibernate, they still have a reduced body temperature (by about 5°C) and their heart rate and respiration drop to save energy, but they still "wake up" to groom, roll over, stretch, stand, and even leave their den for short walks. But like other hibernators, they burn through body weight to survive, and in the Black Bear's case that can be 30 per cent of their weight![31] If you want to see this behaviour on camera, a good winter webcam to watch is the infrared webcam of the hibernation den for the Grouse Mountain Grizzly Bears Coola and Grinder.

One strikingly beautiful, dark-sky location for seeing Black Bears shortly after they emerge from their winter dens is Wells Gray Provincial Park, near Clearwater, BC. It's approximately 5½ hours' drive from Vancouver to Spahats Falls, the first set of falls en route to Wells Gray Park. "Spahats" comes from a local First Nation word for bear, and this waterfall is a great starting point for bear viewing, which you can continue over the next 55 km on a drive all the way to Clearwater Lake. By mid-May, Black Bears are just out of hibernation and are attracted to grazing on the new growth where the forest edge meets the nutrient-rich shrubs, dandelions, clover, and new grasses that line the corridor road. Viewing from a vehicle is necessary as it's the least intrusive and least threatening to a bear, and of course safest for you! Safety, however, is only assured if you also make certain to pull your vehicle well off the road, as there are many blind corners that provide little advance notice to other drivers. It's also critical that you keep a distance of at least 50 m (or three bus lengths), using zoom lenses or binoculars for a closer look, allowing the bears to go about their feeding undisturbed and their behaviour unchanged. Limiting your time viewing the bears to just a few minutes allows the bears more space from humans, because it's likely you will not be the only human observing the bear that day. As you travel the Clearwater Corridor, take your time scanning left and right, watching for bears building their digestive capabilities through late spring to summer, culminating in a diet of up to 20,000 calories/day through August/September as they prepare for next winter's hibernation.[32]

In spring it's possible that when you see a bear it won't just be a lone bear but rather a family of two or three. That family unit will be the mother (sow) and her cubs. A mother Black Bear has just one summer to teach her cubs what they need to learn to survive on their own, and this begins with where and how to find nourishment. In Wells Gray Park, the calorie-packed white and alsike clover that grow profusely along the Clearwater Corridor draw sows and cubs to the roadside, making it an excellent viewing opportunity in spring. As spring progresses, the sow will continue the cubs' education on their daily circuit (they travel the same path most days, occasionally extending into additional areas), showing them not just the *where* and *how* to find calories, but importantly also the *when*. Combining the locational with the temporal (time), the cubs file away strong memories of food sources, which they will rely on in the future. That means it's not uncommon to see the same bear arrive at a particular location (at a salmon-bearing stream, for example), on or about the same calendar day, year after year. This is another reason why careless people leaving food waste at a specific campsite and at a particular time (such as the Labour Day weekend) can have consequences that aren't fully realized until future years, when the bears return unannounced! A bear's positive association between food and humans is a bad association, and since the sow has just one summer to teach her cubs critical lessons, bad habits acquired young can definitely reduce a cub's chances of survival.[33] Ideally, we want our relationship with bears to be neutral, rather than having them associate humans with experiences that are either positive or negative.

By June, it is breeding season and bears without cubs travel farther as they seek out mates, increasing your chance of seeing them in other areas of the park, such as on trails. (Do you have your bear spray handy? Do you know how to use it? And is it on your hip, not in your pack?) And then in July, as the valleys warm and the snow melts higher in the hills, revealing the great expanse of open foraging habitat in the backcountry, the Black Bears travel away from the road and even reduce their movements in order to minimize the heat stress that comes with hot days and a fur coat!

The connection between a mother and her young has often been called the most important relationship in nature. As visitors to their habitat, their home, and their territory, we must understand a sow's need to vigorously defend the space around her cubs, and how easily it can be compromised if we don't fully respect it and do everything we can to protect them.

Wells Gray Provincial Park Highlights

Besides bear viewing, the drive between Spahats Falls and Clearwater Lake leads you to some of BC's most spectacular natural features, including an impressive 39 waterfalls, all within just 55 km![34] Depending on your level of adventure, these include:

→ **Spahats Falls** at 73 m high and the Shaden viewpoint – an easy 2 km walk introducing you to the volcanic history of the park and the sheer cliffs that are common throughout.

→ **Dawson Falls** is next and provides an easy 1 km walk to feel the power of the Murtle River at both the falls and the Mushbowl. The Mushbowl is viewable from a vehicle, but why not get out to feel the immense power of the river?

→ Still farther down the corridor is the highlight of the park, **Helmcken Falls**. It's the fourth-highest waterfall in Canada at 141 m and one of the

SPAHATS FALLS IS THE FIRST MAJOR WATERFALL YOU ARRIVE AT AS YOU ENTER WELLS GRAY PROVINCIAL PARK. OVER TIME THE WATER HAS SLICED THROUGH THE VOLCANIC ROCK! PHOTO STEFAN JÜRGENSEN, FLICKR, CC BY-NC-ND 2.0.

A HIGHLIGHT OF WELLS GRAY PARK, HELMCKEN FALLS LOOKS AS BEAUTIFUL IN THE SUMMER AS IN THE WINTER. ***TRY TO ARRANGE YOUR TIMING SO THAT YOU CAN SEE IT AS YOU LOOK WEST WITH THE SUN AT YOUR BACK AND, IDEALLY, WITH A WIND THAT CAUSES A SPRAY IN THE FALLING WATER.*** **THIS PRODUCES A RAINBOW IN THE WATERFALL MIST.**
PHOTO WILLIAM REDMANN, WIKIMEDIA COMMONS, CC BY 3.0.

main reasons this park was granted protected status in 1939. As with Spahats Falls, you can walk an easy 2 km to view the falls and the Murtle River valley. I'd suggest walking the trail from the viewpoint along the escarpment of the Murtle Plateau to view the confluence of the Murtle and Clearwater rivers. These are both critical in shaping the landscape of the park. You'll be walking on a 200,000-year-old lava field and can almost "read" the layers of lava as you look across the valley at the enormous eroded cliffs. For the more adventurous, the flat 7.8 km return Helmcken Brink Trail (accessed from Dawson Falls) is memorable, allowing you to get close to the top of this awesome waterfall. It's critical, however, to note that the brink of Helmcken Falls is an area *without fencing*, so extreme safety planning is in order. I assure you that you will never forget the experience.

→ The corridor road continues parallel to the Clearwater River and you will have many options and excuses to get out of your vehicle. **Ray Farm** has a natural mineral spring, **Bailey's Chute, Marcus,** and **Myanth waterfalls** all provide easy and powerful nature experiences that focus on the power of spring runoff water. Throughout your journey, watch for tracks, scat, and bear claw marks on the trees to help you explore this habitat more closely. Wells Gray Provincial Park is a geological wonder, and I'd encourage wildlife observers to expand their scope to include geological processes too. Plateau building, lava flows, glaciation, and eroding rivers all have a role to play in this landscape that is laid out in front of you, and, in my opinion, these are the sort of places that can inspire the inner geologist in all of us!

→ **Clearwater Lake**, at the end of the corridor, may be the end of the road, but it is the starting point for a range of other adventures such as a day or overnight ***canoe trip on the lake***. Caribou Beach is about a 6 km return paddle, while Divers Bluff is approximately 18 km return. If you depart early in the day, the afternoon winds may give you a push from behind on your return. Power boat tours venture up to Azure Lake, and from the campground you can do various forest hikes to more waterfalls.

→ To view Black Bears, the suggested time to visit this park is in ***May/June***, but if you happen to choose a different time to visit the park, here are a few more options:

- In summer, the 1017 m **Green Mountain** lookout tower provides an overview of not only the park, its forests, and its fire history but also

FROM LATE JULY THROUGH AUGUST, THE TRAIL UP TO TROPHY MOUNTAIN IS A POPULAR HIKE FOR VIEWING SUBALPINE WILDFLOWERS. ***CARRY AND KNOW HOW TO USE YOUR BEAR SPRAY, THOUGH, AS THE AREA IS ALSO FREQUENTED BY GRIZZLY BEARS FORAGING FOR NUTRITIOUS PLANTS.*** PHOTO JASON HOLLINGER, WIKIMEDIA COMMONS, CC BY 2.0.

the summer weather systems that roll in and affect the landscape. Green Mountain is also the most important winter habitat in the park for Moose. Look for Moose on the mountain from the road just south of the park entrance from ***December to March*** (see the chapter on Moose starting on page 393).

- **Trophy Meadows** are bursting with colourful wildflowers in ***August,*** and these high alpine meadows are feeding grounds for Grizzly Bears as they seek out the bulbs of Glacier Lilies. While in the alpine also watch for hummingbirds, Hoary Marmots, and American Pika.
- **The Horseshoe** and **Bailey's Chute** are great locations to see spawning salmon, including Chinook, in ***early autumn*** as well as many other species feeding on these salmon, including bears, eagles, and others.
- One more thing... ***Spring through fall*** brings approximately 30 unique species of mosquitoes, so be warned and prepare appropriately!

Wells Gray Wildlife Viewing and Recreation Calendar

Human-Wildlife Conflict is a thread that runs throughout this book, and there may be no species where this conflict stands out more profoundly than in our relationship to bears. We have regulations such as timing around putting out our garbage and compost, protocols such as carrying bear spray in the backcountry, and, for some people, emotions that ensure they don't even walk into a local forest due to their extreme fear of bears. Understanding the behaviour of both a bear and yourself in reference to a bear is the best way to keep bears wild and yourself safe. Bears are normally shy of people, but understanding what they do and why they do it are important factors in maintaining safety. There are great resources from WildSafeBC that provide information on "At Risk Campsites," "Encounters with Bears," and "**Wildlife Attractants.**"[35]

The primary attractants are food and waste, and the adage that a "fed bear is a dead bear" is worth considering. Food that attracts bears leads to future expectations for more, and this very quickly leads to food-condi-

BLACK BEARS IN THE CLEARWATER

	JAN	FEB	MAR	APR	MAY	JUN	JUL	AUG	SEP	OCT	NOV	DEC
Cross-country Skiing	Clearwater Corridor											
Moose Viewing	Green Mountain										Clearwater Corridor	
Bear Watching				Clearwater Corridor								
Chinook Salmon Viewing								Bailey's Chute				
Alpine Flowers						Trophy Mnt.	Trophy Mnt.					
Alpine Hiking						Trophy & Battle Mountains						
Lake Canoeing					Clearwater, Azure & Murtle Lake							
Bird Watching				Throughout Wells Gray Park								
Fall Colours									Clearwater Corridor			
Valley Hiking					Throughout Wells Gray Park							

WELLS GRAY PROVINCIAL PARK SUPPLIES THIS HELPFUL VIEWING CALENDAR FOR THOSE WHO WANT TO EXPERIENCE NATURE IN THE PARK IN ALL SEASONS. *CONSIDER THIS IN YOUR PLANNING.*

tioning, in which bears give up their natural food sources and forage from humans instead. Because bears have a strong sense of smell, considerably better than dogs, attractants include far more than a burger on a barbeque. Empty coolers, cans and bottles, plastic food wrappings, toiletries, gas, dog poop bags, and pet food are all signals to a bear that there may be easy calories available. Access to these can lead a wild bear to transform into a problem bear. Problem bears are relocated when possible, but often they must be killed by conservation officers to keep them from returning and harming humans. That's a horrible outcome for all of us, so the responsibility resides on us to minimize these kinds of encounters. But even so, if you're out in bear territory, you may still encounter a bear. Seeing a bear on foot can be an emotional experience, so doing the work to prepare for an encounter is essential if you don't want your emotions to override your logic.

In this chapter I have encouraged the reader to get out and experience Wells Gray Provincial Park beyond the Clearwater Corridor drive, and, potentially, hiking on trails can lead to an encounter with a bear. There are several things you can do to ensure your safety. Here are a few items recommended by WildSafeBC. Hike with others and make your presence known by talking and making noise, while also being aware of your surroundings (e.g., know what fresh bear scat looks like and pay attention to animal-type smells). Dogs in bear country can be an attractant, so leave your furry friends behind if you can, or if you must bring them, keep them on a short leash. Carry bear spray and know how to use it. There are many videos online to help you learn more. Choose a couple from reputable sources and watch them before leaving home. (Every year as spring is around the corner and bears are getting ready to emerge from their dens, my wife and I refresh our bear awareness by doing exactly that.) If you do encounter a bear, remain calm and *don't run,* as it can trigger a bear's chase response. Don't stare or look it in the eye, as in bear language this is a challenge. Talk in a low, calm voice while taking slow steps away, while still facing the bear – speaking ensures that the bear knows you are a human. Don't approach or follow a bear. Consider learning more before your trip, including how to interpret bear behaviour or what extra precautions you can take considering the conditions and geography where you'll be travelling. Bear attacks are rare, but just as you take precautions in other parts of your life, it's worth your time to be prepared in your adventures outdoors.

> To be alive today is to inhabit an Anthropocene world, a world of increasing technological power coupled with ecological illiteracy and a hundred years of indifference to cause and effect. Melting icecaps, acidifying

PAY SPECIAL ATTENTION TO SIGNS THAT ANNOUNCE WILDLIFE IN THE AREA, AND BE SURE TO FOLLOW THEIR DIRECTIONS! **SIGNS LIKE THIS ONE URGE PEOPLE TO REMAIN AT LEAST TEN SCHOOL BUS LENGTHS AWAY FROM BEARS!**

oceans, wildfires, floods, hurricanes, drought, pestilence, extinction: what are these if not Gaia's way of urging us to apprentice to the Living Earth? May the forest be with you.

—**TREVOR GOWARD,** Naturalist, Lichenologist, and Author

Engaging and Taking Action

There are many ways to engage more with bears in British Columbia. Starting from a big picture scope and moving to the personal, I am going to outline several ways you can make a difference to Black Bears in BC. The province has a Bear Smart Community Program so you can advocate for your community to become a Bear Smart Community. The city of Squamish, for example, has designed resources, specifically for their locale, on how to reduce human-wildlife conflicts, including those with bears. Resources are aimed at people who keep bees, chickens, or bird feeders, as well as better ways the community can compost or landscape with wildlife in mind. Squamish also has a video series for hikers and mountain bikers using trails in bear habitat and promotes a wildlife alert system called W.A.R.P. – Wildlife Alert Reporting System.

The BC Wildlife Federation Conservation App is another way to take action, specifically for wildlife infractions. The app allows the user (you can think of yourself as a wildlife warrior protecting wildlife for future generations!) to take geo-referenced, time-stamped photos or videos that relate to issues or abuse that are negatively affecting BC's wildlife, such as people purposefully attracting bears. See the BC Wildlife Federation Map for a more thorough understanding.

You can become a citizen scientist by contributing to the BC Parks iNaturalist Project. Your "observations help researchers develop a deeper understanding of what parks are protecting and can inform management decisions to sustain these protected areas for generations to come."[36] Learn more at the BC Parks Foundation. Begin by creating an iNaturalist account and joining the project.

Finally, another action you can take is to support the creation of protected spaces for biodiversity and, ultimately, the health of our human habitats.

WHILE IN A BC PARK, CONSIDER LEARNING HOW TO USE AND SHARE WHAT YOU SEE ON iNATURALIST, **ESPECIALLY THROUGH THE BC PARKS' iNATURALIST PROGRAM.** ***LEARN MORE ABOUT THIS IN THE INSECT CHAPTER.*** PHOTO DAVID WILLIAMS, FLICKR, CC BY-NC-ND 2.0.

Minimizing Your Carbon Footprint

This chapter has focused on safe bear viewing from a vehicle in the Clearwater Corridor, and therefore a car is encouraged (and under no circumstances should you leave your car if viewing bears). Two carbon-light viewing options are:

1. Bring a friend or fill a car with several friends. Why not make this a social wildlife viewing experience?

2. Wells Gray Provincial Park may offer a good location to consider renting an electric vehicle in Vancouver and then researching the charging stations between the city and Clearwater as you design your wildlife viewing experience. Check out Turo.com or contact the Vancouver Electric Vehicle Association (veva.ca) for their suggestions on this.

If you choose to view bears from the alternate location of Thornton Creek Hatchery (outlined below), then check out the Tofino Bus that runs from Nanaimo or Victoria to Ucluelet. Note that a taxi would be necessary to access Thornton Creek as the hatchery is on the other side of the inlet, 12 km from downtown Ucluelet.

Directions and Accessibility

Take Highway 1 to Hope and then Highway 5 to Kamloops and on to Clearwater. The final 65 km are on Clearwater Valley Road in the corridor to Clearwater Lake and Clearwater campground.

Helmcken, Spahats, and Dawson falls all have viewing platforms that are wheelchair accessible (with rough patches). The dock at Clearwater Lake, a picnic area on the North Thompson, and several campsites also have greater accessibility. For more details contact the Wells Gray Provincial Park Information Centre – www.wellsgraypark.info/; info@wellsgraypark.info; 1-250-674-3334.

Other Locations to View Black Bears

Though BC offers many locations for viewing bears, I'd like to focus on one other in greater detail. Thornton Creek Hatchery in Ucluelet is a remarkable site and a strong alternate location to view Black Bears. Each fall (September to mid-November) the creek provides a predictable, nutrient-rich food source

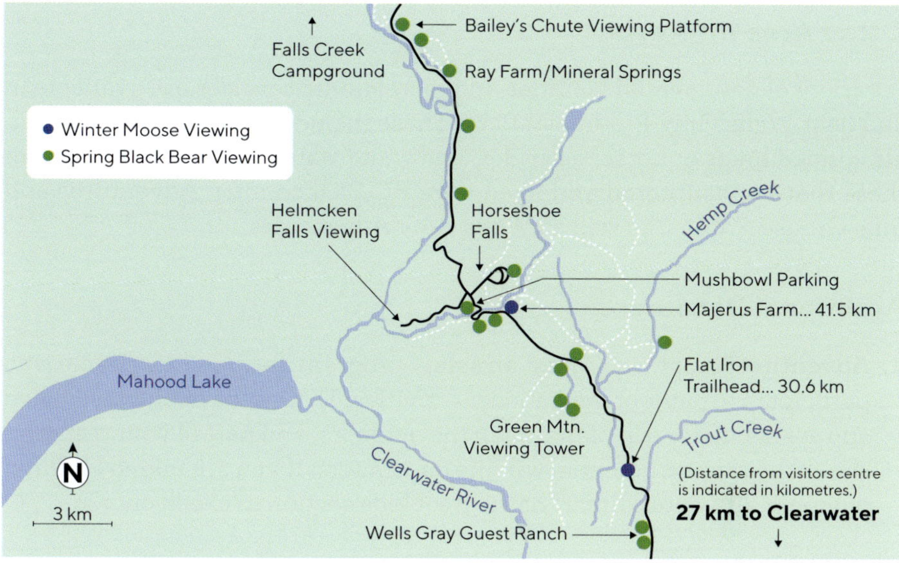

VIEWING LOCATIONS FOR BLACK BEARS (MAY–JUNE) AND MOOSE (FEBRUARY–MARCH) IN THE CLEARWATER CORRIDOR AT WELLS GRAY PROVINCIAL PARK. SOURCE: BC PARKS.

for the bears, consisting of Chum, Coho, and Chinook Salmon returning to their natal streams. Due to its popularity, the hatchery undertook a human-bear conflict hazard assessment, and this led to many measures to ensure the safety of the viewers and to safeguard the bears from impacts. An electric fence was created to "enclose the people" from the bears, as it is often human behaviour that requires controlling (think "Instagram selfie") rather than the bears'. Other critical components implemented by the hatchery were specific hours for viewing (allowing the bears time without people present) and ensuring everyone is escorted from their vehicle into the fencing area, thereby giving volunteers of the hatchery an opportunity to orient and educate 100 per cent of the visitors arriving. This is a very different bear experience than Wells Gray Provincial Park, and if you decide to seek out bears in this location I highly recommend you do some research on "The Salmon Forest" to understand the critical relationship between bears, salmon, and the forests that line many of the watercourses on British Columbia's west coast. For updates, see the Thornton Creek Hatchery Facebook page, but consider visiting during their "BC Goes Wild" weekend in mid-September, as they have other events on at that time too.

Other Area Interests

A variety of commercial facilities and recreational services are available in and near Wells Gray Provincial Park. These include boat tours, canoe rentals, guided hikes or horseback rides, and white-water rafting. A sampling of these that I encountered and liked were BC Backcountry Adventures and Blue Grouse Cabins, as well as Wells Gray Adventures.

Additional Resources

1. **AdventureSmart** is a Canadian safety program designed for people who participate in outdoor recreational activities. Their website and associated app are set up for trip planning in the outdoors, and part of your planning should be around potential wildlife encounters. If you are taking children outdoors, AdventureSmart has a great kids section to check out too.

2. **The Bear Smart Community Program**, WildSafeBC, the BC Wildlife Federation Conservation App, and the BC Parks iNaturalist Project are all great resources discussed above.

CLIMATE CHANGE AND BLACK BEARS

As a hibernating species, bears find the impacts of shorter winters and longer summers, as a result of climate change, more challenging. Salmon and berries are key foods bears eat through the summer and fall to fatten up in preparation for hibernation, but long summer droughts can lead to lower river levels, later salmon runs, and berries drying out on branches before they can be eaten. And if the winter is brief, they then come out of hibernation earlier than usual, burning additional energy in the process.

A UNIQUE NATURAL FEATURE JUST WEST OF OSOYOOS IS SPOTTED LAKE. IT HAS HIGH CONCENTRATIONS OF MAGNESIUM, CALCIUM, AND SODIUM MINERALS, AND WHEN THE SUMMER HEAT EVAPORATES THE WATER, IT REVEALS ITS COLOURFUL MINERAL DEPOSITS. PHOTO MICHAEL SCHMIDT PHOTOGRAPHY.

The Meadowlark Nature Festival

in the South Okanagan & Similkameen

AT A GLANCE

WHAT: Meadowlark Nature Festival
WHERE: South Okanagan – Osoyoos, Penticton, Naramata, Peachland, Summerland, Kaleden, Oliver, Princeton, and other natural locales
WHEN: May long weekend
TRADITIONAL TERRITORY: Osoyoos Indian Band, Sylix Okanagan Nation Territory
ACCESS: Mostly easy but may vary depending on the event

Consider the ways in which you "fertilize" local economies – that is, the ways you "vote" with your dollars to support *local* businesses, organizations, or events. I'd like to highlight the South Okanagan's Meadowlark Nature Festival, which is as far removed from big box stores, corporate conglomerates, and next-day online deliveries as you can get. It is the South Okanagan's biggest wildlife event, but a different kind of big – a small and local kind of big. The Meadowlark festival is one of the largest nature festivals in Canada. Organized by the Okanagan Similkameen Conservation Alliance (OSCA), the festival has been running every May long weekend since 1998 (with the exception of 2020, of course), providing "people of all ages with opportunities to experience, discover, explore, and protect [South Okanagan's] amazing natural environment...in a biologically unique area with both high wildlife diversity and high numbers of species at risk." The festival takes place in an area recognized as one of Canada's most endangered natural systems (which varies between the "unique desert environment in the south, to the alpine meadows high above the Okanagan and Similkameen valleys"), and it provides an opportunity for the public to engage with nature, understand interconnections more deeply, and be inspired by the region and its people.[37] So, what does this look like "on the ground"? The events are led by experts in their respective disciplines such as biologists, botanists, historians, and others. A broad set of subject areas include:

Canoe/kayak trips
Historic hikes
Art exhibitions
Photography
Bird & bat watching
Guided hikes
Writing workshops
Painting workshops
Indigenous culture
Geological tours
Astronomy events
Backcountry on horseback

In addition, there are film screenings, inspirational speakers, and an array of specialists – from biologists and botanists to historians and Indigenous knowledge keepers, and from astronomers and artists to – well, you get the idea. The Meadowlark Nature Festival is extremely rich in its content, appealing to a wide cross-section of people with very diverse interests!

Prices to participate in Meadowlark events vary but are reasonable, with special family packages. But be aware that, due to the festival's popularity and the need to keep group sizes manageable, most events book up well before the festival takes place. Therefore, consider booking your tickets early in April, when registration opens. Each event has a level-of-difficulty rating, so be sure to review these prior to booking. Your registration fee doesn't necessarily include transportation for the events either, since many of the events make use of carpooling (a positively carbon-light option!), so plan accordingly for that as well. The May long weekend can be hot, sunny, windy, cold, or rainy (and sometimes all in the same day!), so dress in layers and bring all-weather clothing with solid footwear. Cameras, binoculars, sun hat, sunglasses, and sunscreen will also add to your experience and well-being, but please keep your pets at home because, although they may have an important place in our lives, they don't always have a place on the tours or in the at-risk natural areas where these events take place.

What does it mean to be an ***at-risk natural area***? The designation indicates that many species are threatened or endangered, while at the same time they are inadequately protected. In fact, the South Okanagan includes the most endangered ecosystems in Canada![38] An expanding human population, along with the associated housing, recreation, and agricultural development that comes with it, creates pressure and stresses on too many species and their habitats. Only 9 per cent of the Okanagan grasslands are relatively undisturbed, while 85 per cent of their wetlands and stream-side habitats have already been lost! Only 3 per cent of the Okanagan basin is

THE YELLOW-BREASTED CHAT IS AN INCREASINGLY RARE SPECIES TO FIND NEAR THE WATER'S EDGE IN THE SOUTH OKANAGAN. THIS BIRD FROM THE WARBLER FAMILY IS AT THE NORTHERN EDGE OF ITS RANGE HERE, AND IS A SIGHT TO SEE AND HEAR IF YOU ARE LUCKY. PHOTO PASCALE CHARLAND.

set aside in parks or reserves, a lower proportion than in other BC regions, such as Metro Vancouver.[39] There are 32 British Columbia designated red- and blue-listed species (**listed species** refer to ranks of a species' conservation status) in the South Okanagan and Lower Similkameen valleys. Listed species that have been outlined in other chapters of this book include the Western Rattlesnake, Painted Turtle, Great Blue Heron, Sandhill Crane, California Bighorn Sheep, and Spotted Bat. One aspirational conservation solution is the United Nations *30 by 30* initiative, which aims to protect 30 per cent of the planet by 2030. As of 2024, Canada's protected areas cover approximately 13.7 per cent of the country (BC protects 19.7 per cent), but our goal is to meet the *30 by 30* target, more than doubling our current protected area. The reasoning behind the 30 per cent goal is clear: we must ensure that the natural areas that provide essential benefits to humanity, such as food, clean water, clean air, and a stable climate, are protected. These are "**ecosystem services**" (see page 226), the benefits that the natural environment provides to us. It's important, however, to ensure that conserving new land and ocean isn't done without the participation of Indigenous Canadians, as many of the today's parks were created from within a colonialist mindset that pushed Indigenous People out of their Traditional Territories. A *30 by 30* goal, though, will go a long way to address the global climate crisis, while also benefiting red-listed species, such as the Okanagan's Yellow-breasted Chat.

The once common Yellow-breasted Chat is now extremely rare. This bird, a member of the warbler family, migrates between Mexico and its breeding grounds in the South Okanagan and Similkameen valleys. It is in danger of becoming extirpated from the Okanagan (that is, wiped out or no longer existing in a specific area) due to habitat loss. Yellow-breasted Chats require the shrub growth found around riparian habitat, and may not be able to survive the loss of wetlands, channelling of waterways, or clearing of shrubs and thickets for agricultural and residential developments.[40] It's likely that a birdwatching event at the Meadowlark Nature Festival will seek out this bird, but if you want to search on your own, consider the following: most Yellow-breasted Chat sightings happen along the Okanagan River, south of McIntyre Bluff and near the north end of Osoyoos Lake. Yellow-breasted Chats sit in dense thickets in riparian habitat and sing early in the breeding season. The males have an extensive repertoire of whistles, rattles, and other sounds. You can listen to their call before your visit at Allaboutbirds.org. Though you may have never heard of the Yellow-breasted Chat, other birds you likely knew well in the past have also decreased significantly in population.

Another well-known bird that is now considered a threatened species across Canada is the Barn Swallow. Although these birds once filled the skies around barns, suburbs, and rural buildings, Barn Swallow numbers have decreased by 75–80 per cent in the last 40 years. Citizen scientists, along with wildlife researchers, have helped to determine the extent of this decline. Seeking out a specific habitat to view one of the South Okanagan's rare species may help you understand at-risk ecosystems better, and perhaps lead you to jump into action – or rather, Seven Simple Actions!

Seven Simple Actions to Help Birds is a website set up by the Cornell Lab of Ornithology. Creating it was a response to a 2019 report that revealed the population of birds in North America had dropped by three billion since 1970! Cornell staff knew that most people love birds and would respond by asking what they could do to help. Their answer was the seven actions.[41] They are:

1. **Make windows safer, day & night** – Day creates reflections, while night lights in cities draw migrating birds toward buildings.

2. **Keep cats indoors** – Other than habitat loss, domestic cats are the number one human-caused reason for the loss of birds in North America.

3. **Reduce lawn grass, and plant native species** – Provide local birds with shelter, rest, and nesting areas.

4. **Avoid pesticides** – Pesticides can harm birds directly if they eat contaminated seeds or prey, or indirectly by reducing the number of insects.

5. **Drink coffee that's good for birds** – Shade-grown coffee preserves a forest canopy that helps migratory birds survive the winter in their southern homes.

6. **Protect our planet from plastic** – Avoid single-use plastics (such as bags, bottles, straws, wraps, and disposable utensils), and if you do use them, be sure to recycle them.

7. **Watch birds and share what you know and see** – Monitor, record, and report birds that you see, and share what you know about birds with others.

CORNELL LAB'S SEVEN SIMPLE ACTIONS TO HELP BIRDS IS AN IMPORTANT WAY TO HELP ALL OF US LEARN HOW TO HELP ALL OF THEM... THE BIRDS, THAT IS! PHOTO THE CORNELL LAB.™

THE MEADOWLARK NATURE FESTIVAL HAS AN ARRAY OF EXPERT-LED BIRD WALKS WHERE YOU WILL LIKELY SEE INTERIOR BIRDS SUCH AS THE PYGMY NUTHATCH, AND MAYBE EVEN LEARN TO RECOGNIZE ITS HIGH-PITCHED PEEP CALLS. PHOTO MICHAEL KLOTZ, FLICKR, CC BY-NC 2.0.

Living in Metro Vancouver, I find that by the first day of March I'm able to open our bedroom window in the evening and awaken to the early season songs of our local birds. Robins, finches, wrens, towhees, and thrushes all begin singing their refrains, and by May, with added migrants from the south such as warblers, vireos, and other birds, it's a veritable dawn chorus! Along with spring's greening of the landscape, birdsong is the best *welcome to spring* I can imagine! Once you learn to appreciate this, you won't be able to imagine not having it! There is something to be said for the sounds of nature, particularly when those sounds are relatively quiet and you have to listen ever so carefully to hear them. Research has shown that natural sounds can be psychologically restorative. There are areas in both the coastal and Okanagan ecosystems that produce a vibrant, organism-rich orchestra of organized expression, and this exists for birds but also for insects, amphibians, and mammals. The idea here is to approach the natural world and appreciate its **soundscape**.

BINOCULARS ARE GREAT, BUT JOINING A WALK WITH A LEADER WHO HAS A SPOTTING SCOPE OPENS YET ANOTHER WINDOW TO THE NATURAL WORLD, ALLOWING YOU TO VIEW WILDLIFE WHILE BEING AT SUCH A DISTANCE THAT YOUR MOVEMENT IS UNLIKELY TO AFFECT WHAT YOU ARE WATCHING. PHOTO GENTRY GEORGE / US FISH AND WILDLIFE SERVICE, WIKIMEDIA COMMONS, CC0 1.0.

THE MEADOWLARK NATURE FESTIVAL OFTEN OFFERS AN OWL TOUR AND A POPULAR BIRD TO SEE ON THIS ARE FLAMMULATED OWLS. PHOTO RICHARD CANNINGS.

There's a wonderful TED Talk, "The Voice of the Natural World" by Bernie Krause, that presents a story of soundscapes and how they can be a valuable tool by which we can evaluate the health of a habitat we're visiting. Bernie breaks a soundscape into three sources: "*geophony*, or the non-biological sounds that occur in any given habitat, like wind in the trees, water in a stream, or waves at the ocean shore...*biophony* is all of the sound that's generated by organisms in a given habitat [such as birdsong or a chorus of toads] at one time and in one place. And the third, is all of the sound that we humans generate and that's called *anthrophony*...Biophonies and geophonies are the signature voices of the natural world, and as we hear them, we're endowed with a sense of place, the true story of the world we live in."[42] Sound is an incredibly important tool – a sense I use to experience the natural world; it widens my scope to 360 degrees, and therefore allows nature to completely envelop me.

There's a great research project in the Hoh Rainforest of Olympic National Park, Washington State, called One Square Inch of Silence. On the Hoh River Trail, about 5 km (a two-hour hike) from the visitor centre, embodying the project name, there is one square inch marked by a small red-coloured stone on top of a moss-covered log. It's not under any flight path, nor near any road, and it is used as a reminder to protect the 'silence' that exists there. The next time you step outside into the forest, stand beside a lake, a stream, the ocean, or walk on a snowy mountain path, consider the soundscapes you know, and where you go to seek 'silence.'

> Every May long weekend since 1998, the Meadowlark Nature Festival has connected people of all ages and interests with nature, engaging and inspiring them to protect the unique habitats and species of this region and indeed our planet. The tours provide a learning experience that demonstrates the importance of Nature in our lives. We are pleased that many of the participants come back year after year to learn more.
>
> —**JANET WILLSON,** chair, Meadowlark Nature Festival

Engaging and Taking Action

Join the Meadowlark Nature Festival email list to ensure you are informed about when tickets to the events are available. This can be done at the Okanagan Similkameen Conservation Alliance (OSCA) site. In addition, if you choose to stay in the Okanagan for an extended length of time, OSCA provides opportunities for people to volunteer during the festival. OSCA also runs the Vaseux Lake Bird Observatory and banding station, which is open daily from August 1 to October 15, from dawn until noon. The public is welcome to visit. Migrant birds are sampled using 14 mist nets, which are designed to catch them gently. They are then banded by experts and released on their way after a very short delay. Visit the Bird Observatory blog for more information.[43]

OSCA also offers environmental education programs called ECOstudies for groups, including school districts, First Nations schools, alternative schools, and home learners, as well as youth and community organizations. If you're affiliated with any of these, consider arranging a field trip.

Another volunteer organization is the Okanagan Similkameen Stewardship, a grassroots environmental non-profit that designs and leads habitat enhancement projects such as native plant enhancement, fencing shorelines, and managing invasive plants. You can check out their upcoming projects at osstewardship.ca/projects.

Minimizing Your Carbon Footprint

The Meadowlark Nature Festival is set up to encourage people to carpool to the events that take place. Therefore, this may be a consideration if you drive to the event. It's approximately 400 km and a five-hour drive to Osoyoos. Mountain Man Mike's Bus Service is another option and their bus travels between the Braid SkyTrain Station in New Westminster and Osoyoos twice a week. Check out their schedule online. The bus service accepts bikes for an extra fee. Consider cycling in the South Okanagan to get to events.

As the festival is set up to attract a wide diversity of ages and interests, consider organizing a family or friends activity, and carpooling together to the South Okanagan.

Travelling Highway 3 in May could be a beautiful drive in an electric vehicle. This might be a good reason to look into an electric car rental. Check out Turo.com or contact the Vancouver Electric Vehicle Association (veva.ca) for their suggestions on this.

Directions and Accessibility

Take Highway 1 to Hope and then Highway 3 past Manning Park, Princeton, Keremeos, and into Osoyoos. Depending on the events you have registered for, you may have to take Highway 97 north toward Penticton or beyond. Check out the Meadowlark Nature Festival website and schedule for details regarding events.

Different events will have different levels of accessibility. Contact the festival organizers for more information.

Other Locations to View Similar Okanagan & Similkameen Wildlife

1. Read the Hummingbirds in Princeton chapter of this book for advice on viewing hummingbirds along the Kettle Valley Railway Trail in May.

2. Read the Small Mammals in Manning Park chapter of this book for suggestions on wildlife in the Manning Park area.

3. Read the Insect chapter, which outlines the dragonflies that may be seen in Cheam Lake Wetlands Regional Park at Exit 138 off Highway 1.

4. Consider a walk in Coquihalla Canyon Provincial Park, through the Othello Tunnels and along the Coquihalla River and gorge.

Other Wildlife Festivals

1. Vancouver Bird Celebration is an early May celebration of Vancouver's birds and coordinated by the Stanley Park Ecology Society.

2. The Tofino Shorebird Festival is run by the Raincoast Education Society and celebrates the spring migration of shorebirds. Lectures led by experts, birding workshops, and educational boat trips are a major part of this late April festival.

3. Parksville's Brant Wildlife Festival celebrates nature in early April, particularly Brants as they stop to feed on the shores of mid-Vancouver Island before continuing to their northern breeding grounds.

4. WildWings Nature & Arts Festival is run by the Somenos Marsh Wildlife Society in the Cowichan Valley on Vancouver Island. The October timing of the festival corresponds with the arrival of Trumpeter Swans, North America's largest waterfowl.

5. Wings Over Water takes place in mid-March in Washington State.

6. Wings Over the Rockies runs in mid-May and is based out of Invermere, BC. This week-long nature celebration includes a diverse array of 100 expert led events and presentations.

Additional Resources

1. ***Wildlife Viewing in the South Okanagan and Similkameen*** is a good brochure to accompany your viewing experience.

2. **The Conservation Pathway Initiative** is about Canada's commitment to protect 30 per cent of Canada's land by 2030.

3. **The San Pancho Bird Observatory** is a wintering home of BC's Yellow-Breasted Chat.

4. **Watch "The Voice of the Natural World,"** a TED Talk by Bernie Krause to learn more about the concept of soundscapes.

5. **Vancouver Co-op Radio at 100.5**, runs the Wetland Project on Earth Day (April 22). They invite listeners to celebrate Earth Day with a 24-Hour Soundscape from a Saturna Island Wetland.

6. Do you like board games? If so, check out the beautifully illustrated **Wingspan**, where players seek to discover and attract the best birds to their network of wildlife preserves.

7. The South Okanagan is a sensitive ecosystem. There are various provincial resources available for understanding sensitive ecosystems in more detail. Check out the **BC Government Sensitive Ecosystem Inventory** website.

CLIMATE CHANGE AND THE SOUTH OKANAGAN

The South Okanagan is one of Canada's most endangered ecosystems. It carries the *at-risk natural area* designation, which indicates that many of its species are threatened or endangered, while at the same time being inadequately protected. Climate change in the Okanagan has led to more droughts, which in turn has led to longer fire seasons, more heavy rainfall events and subsequent flooding, and less predictability of snowpack, which affects freshwater availability.

SUMMER

THERE ARE VARIOUS ORGANIZATIONS DEDICATED TO HELPING PEOPLE WITH MOBILITY CHALLENGES RECREATE OUTDOORS, AND POSSIBLY VIEW WILDLIFE.

INTRODUCTION TO SUMMER

Nature as Teacher

BY DR. RICHARD KOOL

Professor in the School of Environment and Sustainability, Royal Roads University (RRU) where he launched the MA program in environmental education and communication program in 2003. Science teacher in Ucluelet Secondary School, taught at Douglas College in New Westminster, and as a sessional instructor at the University of Victoria and Simon Fraser University. Prior to moving to RRU in 2003, he worked in the education and public programs section of the Royal BC Museum, was head of interpretation in BC Parks, and worked in the biodiversity section of the BC Ministry of Environment.

A summer day at Botanical Beach in Juan de Fuca Provincial Park: lying on your stomach, the sandstone warm under your skin and clothes, and you're looking into a tide pool. The rest of the world simply disappears and you're in a world of Pink Coralline Algae; Purple, Red, and Green Sea Urchins; Six-legged Sea Stars; mesmerizing, swaying brown and green seaweeds; Tidepool Sculpins abruptly swimming and then sinking; Black Chitons, Gumboot Chitons, Keyhole Limpets; and more. So much, all coming into focus unbidden, almost like a hallucination, in front of your eyes.

I've long been a teacher, and as I've gotten older I've come to believe that one of the primary tasks of a teacher is to help *reveal things that are hidden*. Revelation is a dramatic disclosure of something not previously known or realized, and the natural world reveals itself in many ways. In that sense, we are all teachers, all potential revelators, that is, those who are helping others come to these revelations. So, allow me to play with this concept a little. As any good teacher knows, we can't force revelation; we can only open the door to invite the student in. We "welcome them to the club" of those who are already "in the know," and encourage them to learn more and reveal *their* new insights to all of us. Sometimes, it is nature herself who offers open doors for us to enter.

There is no better time for us to pass through nature's doors, as a teacher or a learner, than in the summer. Anyplace, every place, there is nature's invitation as she says, "Open your eyes or close your eyes, activate your ears and your nose, expose your skin. Open up and you'll be amazed and astounded and delighted by what I have to show you, to teach you, to reveal. And if you think you already know everything there is to know, look closer, go in deeper, stay stiller, listen harder, because there is more, and more, and more." These are revelations offered to anyone willing to pay attention, and they are astounding.

Almost anyplace that you'd consider walking in during the summer is, if you know how to look for it, a place of gifts – offering the gift of the time when nature is at its most generous, most luxurious, most inviting, most providing. It can be a full-immersion sensory experience if you're willing to see, to smell, to feel, and to listen. Try this in an alpine meadow, maybe on Forbidden Plateau in Strathcona Park, or en route to Black Tusk in Garibaldi Provincial Park: you're sweating from the exertion, and the Sitka Valerian is in bloom with its strong, delicious scent; so too are the lupine, anemones, arnica, paintbrush, and more, the colours contrasting with the azure sky, while surrounding you is the buzz of pollinating bumblebees, and the persistent calls of our "national bird," those noisy Canada Jays, wishing you'd offer them a token of your appreciation for this beautiful day and for their presence in your world.

One revelation we can experience daily, weekly, or over the course of the entire season, is the extraordinary progression of transformations. Find a pond: you can watch eggs turn into tadpoles, and then into frogs. Stand below a heron rookery: eggs turn into chicks, who later turn into noisy, ungainly, and yet beautiful birds. Look in your garden: in no time at all, eggs turn into caterpillars, then chrysalids, and thereafter into butterflies. We can watch seeds as small as a dot on a page of text turn into straggly seedlings, and then into plants that flower, ultimately making their own seeds. At the ocean's edge: Bull Kelp grows so fast that it seems to almost elongate before our eyes. While at a tide pool: we can see bubbles of pure oxygen covering the water as the green surfgrass, and both the macro- and microscopic algae, carry out the life-affirming, magic-trick transformation of taking nutrients, carbon dioxide, and water and turning them, with the help of the sun's energy, into sugars and oxygen.

Transformations reveal flows of energy. Watch: there's a garter snake grabbing a tadpole; a dragonfly scooping up a mosquito; an Ichneumon Wasp laying its egg into a now-doomed caterpillar. You can see energy flow from predator to prey, from vegetation to herbivore, from detritus to the scavengers that ultimately return molecules back into the dance of life. Nothing in nature is static.

But more than just transformations, there is a world of relationships that are visible if you know where to look. Be careful not to miss the whole world of mutual aid going on all around you. *Mutual Aid* was the title of a book published in 1902 by the Russian anarchist and naturalist Prince Peter Kropotkin, who argued that nature was not primarily about competition between individuals and species, but about cooperation. We see cooperation everywhere: that Keyhole Limpet on the beach will likely have a beautiful worm wrapped around its circumference, hiding under the soft tissue between the foot and the shell; a Graceful Decorator Crab will grab marine sponges and bits of algae and offer them homes on their shells in exchange for providing camouflage from predators; the Giant Green Anemone is green because of microscopic algal cells that make oxygen and sugars through photosynthesis and pass them on to their animal host, who in turn provides both space and nutrients to house and feed the green algal cells. These intimate relationships between species extend from animals to the realm of plants. Dr. Suzanne Simard and her students at the University of British Columbia have shown us the way a range of barely discernible fungal hair-like tubes connect forest trees and shrubs of different species, providing conduits for sharing information and nutrients, one species facilitating the success of another. Those lichens you pass as you hike to the Black Tusk are themselves not one thing but two, a relationship between a fungus that provides a home and a spongy template for microalgae that, as in the Giant Green Anemone, feed the fungus.

Summer is the time for life-changing experiences, a time when it is easy to put ourselves into spaces where we truly know, and feel, that we are part of the natural world, part of the ebb and flow of life, part of the long continuum of humans who have walked, paddled, slept, eaten, and died in the very places we now stand. Look closely at a dragonfly; look into its eye, or into a snake's eye, or a fish's eye; or experience having a hummingbird come up and look at you, eye to eye. Looking into another creature's eye can stop you in your tracks, can shake you to your core; these experiences can be wonderful or even life-changing as you recognize another living creature, not human, is *seeing* you and you have the momentary opportunity to both see and be seen by a wild creature.

We are diurnal creatures and love the daylight; summer is when we have lots of daytime, lots of time to sit quietly and think; sitting quietly almost anyplace for more than ten minutes seems to tell the animals around you that it's okay to come out and be seen and that they can get back to their business. Sitting in nature in the summertime gives us the chance to watch cumulus clouds build up over those areas where the forests are busy pumping moist air into the lower atmosphere, and to notice that in the midday heat, birds tend to be quiet, perhaps resting after the exertion of the dawn chorus and in preparation for what will come as the sun fades in the sky.

And then, as those long days come to an end, we can also enjoy watching the fading of the light into night. It's at this time of day that we see bats hunting whatever they can rustle up in the evening air. Or we might hear wolves or Coyotes howling to each other. And when it's darker still, what could be more wonderful than walking on Long Beach in Pacific Rim National Park and seeing one's footprints glow, or diving into Princess Louisa Inlet in the dark and leaving a glowing bioluminescent track of our body's movements, caused by untold numbers of single-celled creatures named *Noctiluca* generating a few sparks of cold light? Or lying on your back in the pitch dark on a cloudless night after midnight waiting for the Perseid meteor shower in early August; or sitting on the shore of a lake at night to hear the iconic sound of summer: a solitary loon in the darkness.

Summer allows us to be totally engaged, observant, and trustworthy students of the generous teacher who is willing to offer us instruction, no matter where we are, no matter who we're with, no matter the time of day, no matter if we're looking for revelation or not.

Summer can be a time of wonder, a time to indulge our natural instinct of curiosity that may sit somewhat dormant through the other seasons. It's a time, too, when you can, no you *must*, share that wonder with another; take a friend, your child, any child, and get down on your knees to look into a tide pool; smell the mud in the bay; look closely at tree bark; taste licorice root; look closely at a flower bloom; get close to a dragonfly.

Don't let your next summer go by without looking deeply, smelling deeply, sensing deeply, connecting deeply. Pay attention: summer is a time when everything is growing, readying, preparing, transforming. Don't miss the gifts that summer offers you. Keep your eyes open, your senses sharp. You might never have this chance again. The summer has doors to new and rich experiences, and right now they are all open: all you have to do is walk through.

VANCOUVER ISLAND MARMOT PUPS SIZE EACH OTHER UP AS THEY LEARN PLAY BEHAVIOURS CRITICAL TO THEIR DEVELOPMENT.
PHOTO JORDAN CORMACK.

Vancouver Island Marmots

at Mount Washington, Strathcona Provincial Park

AT A GLANCE

WHAT: Vancouver Island Marmots (*Marmota vancouverensis*)
WHERE: Mount Washington Alpine Resort – ski area
WHEN: End of June through July
TRADITIONAL TERRITORY: K'ómoks, Kwakwa̱ka̱'wakw (Kwakiutl), We Wai Kum, We Wai Kai, Coast Salish
ACCESS: Easy to moderate

What's the rarest mammal you've ever seen? Chances are it wasn't rarer than the endemic Vancouver Island Marmot. (Endemic means they're found here, and only here. Nowhere else on Earth!) At the time of writing there are about 303 wild Vancouver Island Marmots living across 20 colonies in Strathcona Park / Mount Washington and the Nanaimo Lakes areas of Vancouver Island, and another 99 in captivity, with some of these ready to be released into the wild next season. However, that's up considerably from an estimated low of just 23 wild and 7 captive marmots in 2003 – a global population of just 30! And in summer 2021, it was especially thrilling to see that a new colony of 10 to 12 new individuals were discovered in Strathcona Provincial Park. However, the fact that there are only 303 wild marmots is why the International Union for Conservation of Nature (IUCN), the global authority on the status of species in the natural world, has the Vancouver Island Marmot on its critically endangered red list of threatened species. That's a lot of pressure for the Vancouver Island Marmot Recovery Foundation (MRF), whose task it is to lead the recovery strategy and the on-the-ground work toward saving this species. Luckily, they had the foresight to start getting the marmots into captivity in 1997 and captive breeding by 2000. Today, their role is to release *captive-bred* marmots and to translocate *wild* marmots. In the process, the foundation monitors known colonies and provides supplemental food in the spring to marmots, while also restoring habitat and reporting out to a recovery team. It's a lot of work, and it is taking a community to save this species.

The MRF plays a central role in a partnership between the recovery team (a team of scientists, researchers, and representatives from provincial, federal, and industry biologists), Mount Washington Alpine Resort, forest management companies, and captive breeding programs at the Toronto and Calgary zoos. This reminds me of the critical nature of Sustainable Development Goal 17, which is often overlooked: Partnerships for the Goals. The Tony Barrett Mount Washington Marmot Recovery Centre is where much of the Recovery Plan work happens and, although the Recovery Centre is not open to the public, captive-born marmots are released and are fairly easy to view in the summer on the ski slopes of Mount Washington. Unfortunately, the low numbers mean the marmots *need* human intervention, and today the species is entirely reliant on recovery action, but it's a long-term goal of the team to ensure the species is self-sustaining, without our involvement. According to Adam Taylor, executive director of the Marmot Recovery Foundation, "We have a moral obligation as this marmot is one of only five endemic species in Canada, and the only one at risk."[44] This is a BC conservation success story in a time where wildlife success stories are critically endangered themselves.

Success is hard won when numbers are so low. Decisions are based on the best available science, but there are still so many unknowns, and you don't always have the luxury of waiting until all the answers are in. One decision, however, proved to be very successful and fortunate for you, the wildlife viewer. Captive-bred marmots, along with some captured wild juvenile marmots, are released every year at varying locations. Often this includes the ski slopes of Mount Washington. Captive-bred marmots from the breeding programs in Calgary or Toronto arrive as pups to the Mount Washington facility in the fall. After a quarantine they spend their first year hibernating on the Mount Washington ski slopes. In their second summer, they are known as yearlings (about 1½ years old) and are given time to put on weight. A stud bookkeeper who tracks their genetics helps decide which marmots will be released. This system has significantly increased the marmot's chances of survival. What was a 1 in 20 chance of survival (to breeding age), when marmots were first released from captivity into Strathcona Provincial Park in 2007, is now closer to 1 in 6! The key reasons are that human activity is thought to discourage predation at Mount Washington, and that the mountain has abundant burrows that support captive-release marmots in their first hibernation. Thus, after a year on Mount Washington they are much better candidates for translocating to other areas in Strathcona

Park or Nanaimo Lakes. The learning and growth that take place enables a *fivefold survival rate,* getting them closer to breeding age. People, in this very public space, keep predators of the marmots at a distance because the predators are much more wary of hunting a marmot on the open ski slopes with people nearby. Eventually, when the marmots are released into more remote areas, they are more adept in how they respond to Grey Wolves, Cougars, and Golden Eagles.

The idea is to build the population now, so they are more resilient to future harmful events. When declines happen in future years – and they will – they'll likely come without warning. One extraordinarily long winter with deep snow, or a particularly lengthy drought that leads to wildfire, can have a disproportionately devastating impact on the species. I asked Dr. Malcolm McAdie (an MRF contractor who happens also to be a wildlife veterinarian) for a memorable story from his 30 years of working with Vancouver Island Marmots. He told me about how it felt in 2003 when they first released four captive-bred marmots to Green Mountain. Remember, there were only 30 Vancouver Island Marmots left in the wild at that time. They monitored the released marmots daily, but still three were predated on by a Cougar. So they decided to recapture the final one while the opportunistic Cougar looked on from the bushes! I could hear the stress in Malcolm's voice when he retold the story. He continued, "So what do you do? Do you keep going? You *have to*. You don't have a choice." The goals were broken into small steps at the time, such as just getting them through the summer, then through hibernation, and then to breed. As they learned, their decisions eventually led to success, and by 2012 the marmot numbers were consistently increasing. Due to this early work, it is possible today to see this species that almost, in Malcolm's words, "winked out."[45]

Viewing Vancouver Island Marmots at Mount Washington requires respectful viewing practices. July is a great time because the pups are moving about. The ski slopes of Mount Washington, it turns out, look exactly like marmot habitat! Marmots like small patches of south- and west-facing subalpine and alpine meadows (ideally above 1000 m), avalanche chutes, and fields with boulders for perching on. They choose areas that become snow free, leading to food earlier in the season. If you see them, keep at least 100 m away (that's eight school buses) and watch their behaviour. If they stand up straight or whistle, it may be a defensive behaviour, indicating they're feeling threatened. If they run to a burrow and hide, then it's very likely a protective action, so back away. Please don't try to get their attention, and certainly keep your food in your pack. Also, consider viewing from the chair lift. If your visit is on a warm summer day, it's best to view early or late in the day, as the marmots aren't very heat tolerant – because they

don't pant like dogs or sweat like us to release excess heat. Instead, they move to their underground burrows to keep cool. Transfer of diseases from other areas is also an issue for the marmots, so before arriving at Mount Washington *please clean your dirty footwear and gear*. As with other wildlife viewing, sightings are not guaranteed, but patience and time will help.

The marmots you're looking for are cat-sized and generally chocolate-brown. They are a member of the squirrel family. When you see one, it's likely you'll agree they're unbearably cute, endearing, and entertaining to watch. They have an array of behaviours, such as lounging or scanning for predators while perched on boulders, foraging for food, greeting each other with nose touching, and play fighting (pup boxing). Adam Taylor calls them an "ambassador" with the ability to hook people into caring about the conservation of species at risk. Marmot mothers usually have 3–4 pups in a litter, and since marmots can live to over ten years in the wild, if a female remains healthy, she can produce 12–15 pups in her lifetime. In 2023, 53 pups were born in the wild and 42 were released. Sixty-five marmots are scheduled to be released in 2024. That's great news for the species![46] But with a population that was as low as 30, you can imagine how narrow their genetic diversity is. High genetic diversity results from a wide gene pool, and generally results in healthier individuals and a more resilient population. Inbreeding within small populations, on the other hand, tends to result in poorer health outcomes (for example, hip dysplasia is a common affliction in certain breeds of dogs) and lowers the ability of the species to adapt to changing environments.

A term connected to low genetic diversity is the idea of a **genetic bottleneck**. This means that just one event, such as disease, can have a catastrophic, domino-effect impact on a species, because all of its members have similar genetics when there are no populations from outside areas to add diversity to the gene pool. Climate change will certainly throw some unknowns the marmots' way, particularly in terms of a changing habitat. But since captured marmots have been captive-bred since 1997, the recovery team (which, as mentioned earlier, includes a *stud bookkeeper*) has been considering this for breeding decisions since that time, aiming to ensure the widest possible diversity in the population and also ensuring that marmots born with congenital health problems don't breed. Today, all captive-bred and a few wild marmots have been surgically implanted with devices to monitor their locations. The health of both captive and wild marmots is monitored through regular physical examinations that include bloodwork and checks for parasites and other related health issues. This is especially important to help them get through the critical period of hibernation.

Hibernation is a metabolic process that comes up repeatedly throughout this book. Whether you are observing rattlesnakes in Osoyoos, Painted Turtles in the Fraser Valley, Black Bears in Wells Gray Park, bats in Peachland, or ground squirrels in Manning Park, hibernation is part of *all* of their annual cycles. And when a species hibernates as long as the Vancouver Island Marmot (a phenomenally long seven months!), you can appreciate how everything they do while awake must revolve around this cycle. That means they need food such as grasses and sedges to be available when they tunnel through snow to open their burrows in late April, and they thrive when the summer plants and flowers grow lush and are prolific, allowing them to build up required nutrients before fall. It also means they benefit when winter snows arrive at a time that coincides with the decay of vegetation, allowing them to enter hibernation with sufficient nutrients and fat stores. The truth is, over winter, mortality from starvation is a real issue for these marmots.

Consider what hibernation looks like for marmots. First, they descend into their hibernacula (underground dens) to hibernate together in groups of up to six. Their heart rates drop from an active 200 beats/min. to just 3–4 beats/min., respiration happens just once each minute, they lose up to one-third of their body weight, and their body temperatures (normally slightly warmer than ours at 38°C) drop to just 5°C. Even their brainwaves become virtually undetectable. But hibernation itself is also cyclic. About every two weeks, they warm up together in a physiological cycle that brings their brainwaves into a state comparable to our REM sleep, and increase their temperature, respiration, and heart rate, and then they cycle back down, again all together. Why is this? Though we don't have all the answers, Dr. Malcolm McAdie thinks it's about rebuilding tissues, including brain matter. Considering 80 per cent of their energy is expended during the arousal portions of their hibernation over this seven-month physiological cycling process, it's imperative that they go into their hibernacula below the frost-line and in the best condition possible. Malcolm is responsible for surgically implanting radio transmitters that send a pulse to help researchers monitor the marmots during their summer movements and their winter hibernation. Pulses are temperature responsive, and this, in turn, provides an assessment of their health. This is *learning in real time*. "There are still many questions we just don't know, but the more we learn," according to Malcolm, "the better our next questions become."

Learning in real time. Think about that. Our learning about this species is only decades old, beginning with the research of Douglas Heard in the 1970s. That's just a moment in time for a self-sustaining species that has seen glaciers advance and retreat, forests ebb, flow, and change, and alpine meadows freeze, thaw, expand, and contract – over millennia. However, sadly they are no longer self-sustaining, not without our help. As humans, we are changing the regularity of these environmental cycles as we alter the climate. If you see marmots looking over their shoulder, it may be the climate they're looking at; that is, they may be looking at change in the form of an especially rapid spring melt, or an unusually dry summer slope. How will climate change affect their habitat? We're certain warmer temperatures

SUBALPINE MEADOWS ARE THE BEST FORAGING GROUNDS FOR MARMOTS, AND THEY SUPPLY THE BULK OF THE CALORIES REQUIRED TO LAST THE MARMOTS THROUGH THEIR LONG WINTER HIBERNATION. PHOTO CHRISTINE GAIO.

VETERINARIAN MALCOLM MCADIE WORKS WITH AN ADULT MARMOT AS HE PREPARES TO SURGICALLY IMPLANT A RADIO TRANSMITTER TO HELP RESEARCHERS MONITOR THE MARMOT DURING ITS SUMMER MOVEMENTS AND WINTER HIBERNATION. PHOTO MIKE LESTER.

will lead to tree cover expanding upslope and choking out existing meadows, but ultimately the problem of protecting Vancouver Island Marmot habitat is a baton Malcolm and Adam will be handing off to others in the next generation to solve.

In the meantime, we can reflect on what we value. Let's consider a few value-based items. A top-scoring defenceman for the Vancouver Canucks with an annual salary in excess of $10 million. A beautiful Group of Seven painting titled *The Old Stump, Lake Superior* auctioned for $3.5 million. An elite luxury car, a colossal 7,500-square-foot house, a bigger boat…you get the idea. If only the Vancouver Island Marmot Recovery Team had such

IT TAKES A TEAM TO ENSURE THE RECOVERY SUCCESS OF VANCOUVER ISLAND MARMOTS. THE VANCOUVER ISLAND MARMOT RECOVERY FOUNDATION LEADS THE RECOVERY STRATEGY AND THE ON-THE-GROUND WORK DIRECTED TOWARD SAVING THIS SPECIES. PHOTO JORDAN CORMACK.

resources. They have their hat out asking for donations every year and must justify the cost of everything they do. How much of our limited resources should we put into saving a species? What's the Vancouver Island Marmot worth to us? While you're thinking about that, consider these two ideas. The burrows marmots make are used by many other species, including toads, snakes, and pollinators, and the soil they turn up plays an important role in the cycling of nutrients for plants in an otherwise harsh ecosystem. Furthermore, the Vancouver Island Marmot itself is host to a myriad of other species (tiny and micro-organisms, unseen by the naked eye), such as a particular roundworm found nowhere else but on them, which depend on the marmots for their own survival. If we drive the Vancouver Island Marmot to extinction, other species may be extinguished as well.

That means public support of breeding recovery efforts is critical. Malcolm mentioned that gravel backroads are a favourite place for marmots to burrow as they are easy to dig in. It turns out, however, that old resource roads are also well travelled by two of their predators – Grey Wolves and Cougars. After their second hibernation with their family, juvenile marmots leave their natal colony to find a mate and a new territory to call their own. This is known as **dispersal**, and if you consider the genetic bottleneck discussed previously, you can understand the importance of dispersal. Dispersing across a landscape leads to slightly different habitats that a marmot has to adapt to, and in those adaptations they increase their resilience to external influences. The young marmots typically travel between 5 and 20 km, but can travel even farther. Over the years, after leaving mountainous areas they were released in, they've turned up in Bamfield, Nanoose Bay, Whiskey Creek, and even Jingle Pot Road in Nanaimo! Those spotted in poor marmot habitat are picked up and relocated by the Marmot Recovery Foundation. Therefore, it's important to note that they don't just require suitable habitat, but also travel corridors allowing viable links to other healthy marmot colonies, corridors that are not riddled with roads or widespread industrial activity. As it says on the foundation website at marmots.org, dispersal "guards against a total collapse of the species if any one colony should suffer a catastrophic event." Just like our grandmothers told us, "it isn't wise to keep all your eggs in one basket."[47]

> I started by helping out in the Vancouver Island Marmot program through volunteering in the 1990s. I can see the changes in this history, but that's really only a moment in time! When you consider how the glaciers' expansion and reduction have affected the forests and meadows and the marmots for millennia, you really get a sense of the animal's survival over time. At the moment, we are only talking about the ups and downs of the last 30 years!
>
> —**DR. MALCOLM MCADIE,** veterinarian

Engaging and Taking Action

Perhaps the best way to take action for Vancouver Island Marmots is to learn more about this marmot story and share it widely. Creating general awareness among the public helps build support for its recovery. Many people, even Vancouver Islanders, are still unaware the species exists, or believe that not much can be done as the species is already on its way to extinction. We don't have to let that happen!

You can donate through the Adopt-a-Marmot program. You can report sightings to the Marmot Recovery Foundation if you see one while hiking in the backcountry. You can remember to brush off the soles of your boots and spray them with diluted bleach to ensure you're not transferring a potential disease to marmot habitat. You can keep your dogs on a leash when you're in marmot habitat and leave them at home, if possible, when you visit Mount Washington.

Minimizing Your Carbon Footprint

Mount Washington Alpine Resort has options for shuttle services that connect with Courtenay or Nanaimo. IslandLink Bus runs between Departure Bay Ferry Terminal in Nanaimo and downtown Courtenay. It's 33 km and 1200 m in elevation between Courtenay and Mount Washington Alpine Resort, but for those who like to cycle uphill, it's a lush ride to the resort, with summer-blooming shrubs and flowers bordering the road all the way to the top. The roadway is excellent and there's a paved shoulder most of the way. As with any mountain climb, though, be extra careful on the descent if you choose this carbon-light way to visit the marmots.

Directions and Accessibility

If you begin at either the Horseshoe Bay or Tsawwassen ferry terminal, go north on Island Highway 19. Go left at Strathcona Parkway / Mount Washington Road just north of Courtenay.

Mount Washington Alpine Resort has areas that are fully accessible. Depending on where the marmots are in relation to the resort, they may be viewable from a fully accessible area or not. Contact the resort to ask if they are viewable from the Mount Washington accessible alpine area.

Other Locations to View Vancouver Island Marmots

1. Mount Arrowsmith Regional Park – Below the summit on both sides of Arrowsmith are marmot colonies. It is worth noting that Mount Arrowsmith is part of a wider UNESCO Biosphere Region. British Columbia currently has three biosphere reserves: Clayoquot, Mount Arrowsmith, and Howe Sound. Check out all three at biospherecanada.ca/amazingplaces.

WHAT WAS A 1 IN 20 CHANCE OF SURVIVAL TO BREEDING AGE, WHEN MARMOTS WERE FIRST RELEASED FROM CAPTIVITY INTO STRATHCONA PROVINCIAL PARK IN 2007, IS NOW CLOSER TO 1 IN 6 DUE TO LESSONS LEARNED SINCE 2003.
PHOTO JORDAN CORMACK.

2. Colonies exist in the backcountry of Strathcona Provincial Park in the following areas: Marble Meadows, Tibetan, Greig Ridge, Albert Edward at Hope Lake, Frink, Castlecrag, George V, and Flower Ridge.

3. Nanaimo Lakes area, as well as Haley Lake Ecological Reserve, where access is restricted and permission required.

4. Both the Toronto Zoo and Calgary Zoo have marmot breeding programs, but the facilities are not open for viewing by the public. Handling marmots is minimized, and they are kept in outdoor pens that simulate their natural habitat. *There is NO marmot viewing at either zoo.*

5. The Tony Barrett Mount Washington Marmot Recovery Centre also has a captive breeding program, and it is there that marmots are acclimated to release conditions with respect to weather, elevation, and natural foods. *There is NO marmot viewing at the centre.*

Other Area Interests

1. The Courtenay Museum and Paleontology Centre has a great paleontological exhibit and fossil tours that "send you" 80 million years back in time to learn about the Puntledge River Elasmosaur, a ten-metre dinosaur from the Late Cretaceous period.

2. Strathcona Provincial Park has many opportunities to hike, cycle, canoe, and nature watch.

Additional Resources

1. ***Waking up from Hibernation*** – a 2½-min. BBC video that helps the viewer understand the process of a marmot waking up from hibernation in the spring.

2. **Marmots.org** is the Marmot Recovery Foundation's website. This is the "go to" place for learning more, and about the latest developments on the Vancouver Island Marmot.

3. **The Sustainable Development Goals** (SDGs) Progress Report – This annual report includes a graphic to help the reader understand where their country stands in their commitments to the SDGs. Check to see how Canada is doing relative to other countries in protecting biodiversity.

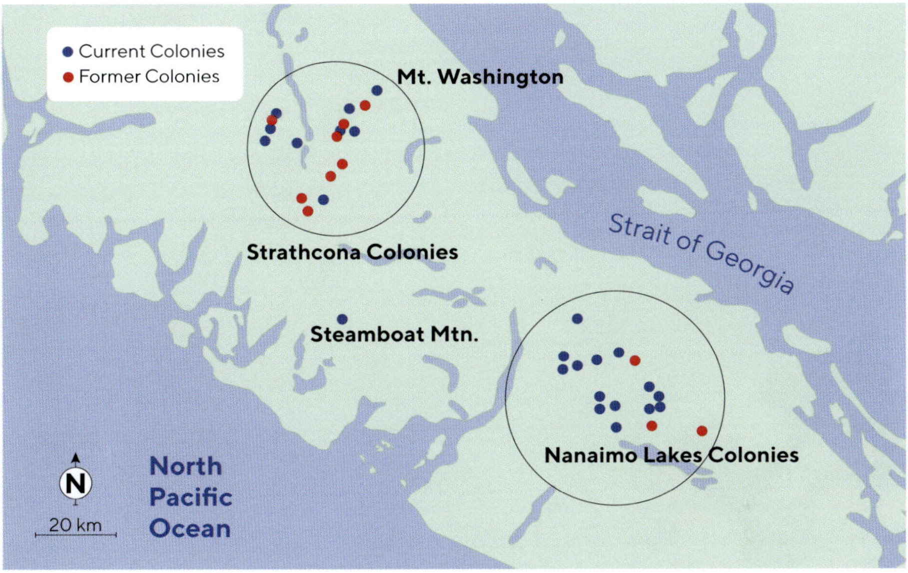

VANCOUVER ISLAND MARMOTS ARE PRIMARILY FOUND IN TWO GEOGRAPHIC AREAS AT MOUNT WASHINGTON, STRATHCONA PROVINCIAL PARK, AND THE NANAIMO LAKES AREA.
SOURCE: MARMOTS.ORG.

CLIMATE CHANGE AND VANCOUVER ISLAND MARMOTS

We know the limited genetic diversity of Vancouver Island Marmots doesn't allow much leeway for major deleterious events, and climate change will certainly throw some unknowns the marmots' way, particularly in terms of changes to their habitat. Warmer temperatures will lead to tree cover expanding upslope and choking out existing meadows used by the marmots. Unseasonable mid-winter thaws can melt the snow, then refreeze as a ground level ice layer, potentially creating adverse consequences for species hibernating underground.

"GARDENING FOR BUTTERFLIES" IN OUR COMMUNITIES CAN BRING BEAUTIFUL INSECTS, INCLUDING THE PAINTED LADY BUTTERFLY, CLOSER TO HOME, WHERE WE CAN MORE EASILY WATCH THEM. THERE ARE NUMEROUS RESOURCES AVAILABLE TO LEARN MORE ABOUT THIS INCREASINGLY POPULAR VIEWING ACTIVITY.
PHOTO MARION SHIKAZE.

Insects: Butterflies, Dragonflies, Damselflies, & Bumblebees

in Skagit Valley Provincial Park & Cheam Lake Wetlands Regional Park

AT A GLANCE

WHAT AND WHERE: Butterflies (*Lepidoptera*) in Skagit Valley; Dragonflies (*Odonata*) & Damselflies at Cheam Lake Wetlands Regional Park; Bumblebees (*Hymenoptera*) in a field of flowers near you!

WHEN: Butterflies: April–September; many peak in June especially on summer days after two or three days of sun. Dragonflies & Damselflies: Sunny days (11 a.m.–4 p.m.) from late May through October. Bumblebees: April to autumn.

TRADITIONAL TERRITORY: Stó:lō, Coast Salish

ACCESS: Skagit (moderate due to 60 km gravel road); Cheam Lake (easy); Bumblebees (easy)

Insects are discussed throughout this book. That's because they're foundational to well-functioning ecosystems and are an essential connection to many other species. In Princeton, a baby Rufous Hummingbird is grown on "mostly soft and fatty, smashed insects" (according to hummingbird researcher Alison Moran). At the Peachland maternity roost, a female Little Brown Bat feeds on up to 1,000 insects per hour, or 110 per cent of its body weight in insects *each night!* At Goldstream Provincial Park, a salmon fry grows into a smolt before making its way to the ocean, only because it's been able to ingest an abundance of insects and their larvae in the river. At Cheam Lake Wetlands Park, an agile Canada Darner Dragonfly uses hunting skills honed over millions of years to capture insects and reign supreme around its watery world. These are just a few examples of how so many wildlife species are *directly dependent* on insects. When you add *the indirect connections*, you could make an argument that every species outlined in this book relies on insects in some way to flourish. Killer Whales eat Chinook Salmon (and what do young salmon eat?); Great Blue Herons eat amphibians (and what do amphibians eat?); Vancouver Island Marmots eat flowers and grasses (and what pollinates these?). Bald Eagles and Black Bears, Harbour Seals, and Red-tailed Hawks – they all rely, directly and indirectly, on healthy insect populations in order to thrive, themselves. Unfortunately, insects don't factor largely into our conservation efforts. In fact, instead of protecting insects, we find a multitude of ways to reduce their populations to suit our needs, including aerial spraying of bacterial insecticides for spruce budworm, neonicotinoid pesticides on blueberry fields, and bacterial larvicide to control mosquitoes in populated areas.

To learn more about insects, be sure to read the chapters in this book that explain concepts such as the *windshield phenomenon* (bats in Peachland) and *"neonic" pesticides* (hummingbirds in Princeton). The remainder of this chapter will focus on three specific insects: butterflies, dragonflies, and bumblebees. I'd like to propose an idea for the growing naturalist in you – to photograph, identify, and post to iNaturalist (see Engaging and Taking Action below) a *minimum of three* species from each category this summer.

Butterflies

If I asked you what comes to mind when I say the word "insect," do you form a positive or negative association? Perhaps you went negative with mosquito bites or a bee sting; or did you go positive with a past butterfly experience? The truth is, our relationship to the natural world is complex: we see both beauty and danger in it and insects can elicit both responses. **Complexity** is such an important idea in nature. Butterflies vary in size, shape, and colour and can be found in a wide variety of BC habitats that range from our backyard gardens to coastal bogs, from alpine meadows to the Okanagan pocket desert. The reason for this diversity is that certain butterflies have adapted to have particular affiliations with specific plant species, and the plants themselves range widely. Crispin Guppy co-authored a book titled *Butterflies of British Columbia*, and in it there's a chapter dedicated to the impact of humans on the butterfly fauna of BC. The chapter reflects on the 19th-century opening of the West to European settlement. It discusses *five major impacts*, starting first with "the alienation of natural habitat by increased housing, industrial sites, roads, hydroelectric dams, recreational sites, and farmland...especially on southeastern Vancouver Island, the Fraser Valley and southern Okanagan." My immediate thought was that this sounds similar to issues that are still relevant today. The *second* set of impacts included the introduction of weedy plants such as grasses that in turn replaced nectar sources (flowering plants) and larval food plants (plants that are fed upon by the larvae of specific insects, such as butterflies). The *third* impact has been as a result of fire suppression. It turns out that protecting and growing more forests may be great for capturing carbon, but mature forests are unsuitable habitat for most butterfly species. *Fourth* and *fifth* are the introduction of non-native butterfly species, such as the Cabbage White, and the use of pesticides and the spraying of biological control agents, which have led to drastic decreases in the populations of ALL species in the order of Lepidoptera (butterflies and moths).[48] I had the privilege of speaking with Crispin (Cris), and he reminded me that his book also has chapters on conservation efforts, including gardening for butterflies.

Cris told me there isn't a single native butterfly species population in BC that is increasing (they are all either stable or in decline), and that in order to turn that around we should be focusing on the protection of our grasslands, wetlands, and sandy-gravelly areas that support native plants. He said the big difference will come when we're able to scale up the production of factory/lab-grown meat, the result of which will be the removal of cattle from grasslands (which have been transformed from their native state, and are now covered in non-native grasses), allowing the restoration of that habitat to nectar-producing and larval food plants.[49] As individual BC residents, we make choices in our communities: we can make decisions to grow larval food plants in our yards; introduce butterflies into elementary school classrooms; join butterfly counts; and generally become more understanding of the plight of these beautiful insects.

If you have space around your home or see opportunities to plant for butterflies in your neighbourhoods (e.g., community gardens), then you should know about The Butterflyway Project. Winnie Hwo leads this project for the David Suzuki Foundation, and she told me about their Ranger program, which exists across Canada, including in many Metro Vancouver neighbourhoods. The "Ranger" is a community leader who facilitates many others to plant gardens that flourish for butterflies and other insects. The most important factor for a Ranger is not that you know your butterflies or plants necessarily, but that you have the ability to motivate other community members to adopt the idea![50] The City of Vancouver also has the Green Streets Program, and all Metro Vancouver communities are increasing community garden opportunities for their residents. Butterflies can often be found right

ANISE TIGER SWALLOWTAIL AND OTHER BUTTERFLIES HAVE ADAPTED TO HAVE PARTICULAR AFFILIATIONS WITH SPECIFIC PLANT SPECIES, AND THEREFORE IT'S HELPFUL WHEN LOOKING FOR A BUTTERFLY TO KNOW WHICH PLANTS IT PREFERS.
PHOTO JOHNATHAN MWENIFUMBO.

in our own gardens and communities, but if you're interested in exploring a natural landscape this summer, the areas of Skagit and Manning provincial parks are ideal places for butterfly watching.

Both Manning and Skagit Valley provincial parks are in the North Cascades, a rain shadow area with less frequent summer rains, and if there's one thing a delicate butterfly seeks, it's little rain and lots of sun. In fact, butterfly watching is best on days after the sun has shone for several days in a row. The Skagit Valley offers up low elevation flora with associated butterflies, while Manning's high elevations provide a completely different pairing. Denis Knopp, a Fraser Valley endangered species specialist and butterfly enthusiast, told me butterflies are a "nine to five" species (referring to the time of day you'll likely find them). Beginning in late May and throughout June, you can locate and identify multiple species along the Skagit Valley Road between the Silvertip campground and the day use parking areas of Shawatum, Rhododendron Bar, Strawberry Bar (especially parking spaces with puddles where butterflies seek minerals), as well as Whitworth and Chittenden meadows (including the 3 km loop trail to old Ponderosa Pines), and the Ross Lake campground.[51] At the higher elevations of Manning park, adult butterflies appear in mid-June and fly through until mid-August. Check out the park headquarters area, the road to Blackwall Peak, and the meadows that border the road to the top. The meadows of the upper Skyline trail are also great places to look.

Representative butterflies of the Skagit Valley peak in June (three days of sun after rain):

Western Tiger Swallowtail (April–September when azaleas bloom)
Mourning Cloak (March–October when willow is present)
Lorquin's Admiral (May–September, with willows and cottonwoods, and in cooler areas)
Margined White (April–September, when mustard family plants are in bloom)
Woodland Skipper (July–August, when goldenrod is in bloom)
Clodius Parnassian (May–September, lays eggs on Bleeding Heart)
Sara's or Stella's Orangetip (March–July, when Erabus family of flowers are in bloom)
Satire Anglewing (February–October, Stinging Nettle)
Green Comma (April–October, willows, alder, and birch)
Boisduval's Blue (April–August, when lupine is in bloom)

1. WESTERN TIGER SWALLOWTAIL. PHOTO TERRY CARR.
2. MOURNING CLOAK. PHOTO CHRIS MCVITTIE.
3. LORQUIN'S ADMIRAL. PHOTO CHRIS MCVITTIE.
4. MARGINED WHITE. PHOTO MARION SHIKAZE.
5. WOODLAND SKIPPER. PHOTO MARION SHIKAZE.
6. CLODIUS PARNASSIAN (FEMALE). PHOTO ROB ALEXANDER.
7. SARA ORANGETIP. PHOTO BILL RAMEY.

Dragonflies and Damselflies

Dragonflies and damselflies, the order Odonata, are mainly a tropical group, and therefore, in British Columbia, you are most likely to find them in areas with warm water. Unlike with birds, where you have to get up early to hear their dawn chorus in spring, dragonflies are most likely to be seen during the hottest and sunniest times on summer days, so early rising is not necessary. If you spend time watching them, certain species may perch near you, hover in front of you, or even land on you! While in flight, many males vigorously defend their territories, while others patrol for females. All dragonflies begin their lives in a pond as larvae (also known as **nymphs**), and many take about a year (though some can take three or more years) to mature before we see them in flight. So, consider looking for them in their larval stage too. Observe their agility in flying and their iridescent colours and, if you get close enough, try to make out their compound eyes. Distinguish the three distinct segments of their bodies, common to all insects: the head, thorax, and abdomen. If your timing is right, you may also see a pair mating, in which the male grasps the female at the back of the head (and at

WATCH FOR A MATING PAIR OF NORTHERN OR BOREAL BLUETS. THE MALE GRASPS THE FEMALE AT THE BACK OF THE HEAD WHILE THE FEMALE CURLS HER ABDOMEN FORMING WHAT MAY LOOK LIKE THE SHAPE OF A HEART. PHOTO TERRY CARR.

the front of the thorax in damselflies), while the female curls her abdomen around to pick up sperm, their contortions forming what may look like the shape of a heart.

The presence of dragonflies indicates a healthy water ecosystem, and although you'll often see them around all sorts of ponds, it's those ponds that have been around the longest (the "old growth" of ponds) that harbour the greatest diversity of dragonflies, according to Sydney Cannings, a species at risk biologist for Environment and Climate Change Canada.[52] Old ponds have had the time to mature through a succession of vegetation changes that result in a diversity of habitat requirements that dragonflies seek. We may see ponds as lily pads and cattails, or sedges and rushes, but dragonflies see them as home; and just as we have certain preferences for our home, yard, neighbourhood, and neighbours, so do insects such as dragonflies. If you have peatlands near you, they are critical dragonfly habitat, so don't let them go. These, along with other wetland environments, are critical for this order of insects, which generally only exist as aerial predators (when we typically notice them) for about 1–2 months but use the habitat for years in the other stages of their life cycle. If you do spend time watching dragonflies, consider their evolutionary history, which is recorded as fossils dating back 325 million years![53] The ponds and water ecosystems described here may match many locations in Metro Vancouver and the Fraser Valley, but for this chapter I'd like to lead you to a beautiful wetland returned to nature, from what was a mineral quarry.

Cheam Lake Wetlands Regional Park is a great example of ecological restoration that has increased biodiversity in this area. The park has several kilometres of walking trails, a boardwalk, and a viewing platform, all surrounded by a lake, marsh, and forest habitat. Take the 1.3 km loop to the left of the viewing platform at the parking area to walk through this sensitive habitat that's home to an abundance of insects, plants, and birds. Another option in certain wetlands or lakes is to get on the water with a stand-up paddleboard or paddle a canoe or kayak. These offer the ability to move slowly and to pause while watching insects in their territories. But only consider this if you are certain you will not be disturbing nesting birds. Always check the posted signage of the area you are in.

To prepare you for success on your dragonfly and damselfly walk, I'd like to introduce you to several online tools. With an iNaturalist account you can view a checklist of species that have been recorded at Cheam Lake Wetlands. Use the site to preview the species you may see. Another fantastic resource is the electronic atlas called E-Fauna BC. It's an atlas of wildlife in British Columbia and it covers the natural history of every species in this book, along with locations they inhabit in the province. Housed at UBC,

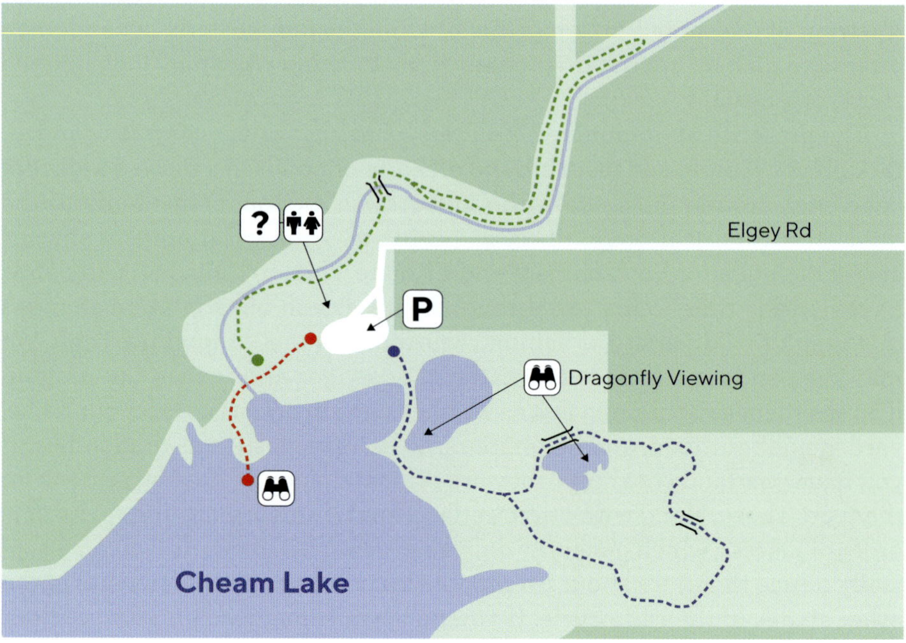

CHEAM LAKE WETLANDS REGIONAL PARK IS A GREAT EXAMPLE OF ECOLOGICAL RESTORATION THAT HAS RESULTED IN INCREASED BIODIVERSITY IN THIS AREA. THE LAKE AND MARSH ARE PARTICULARLY GOOD FOR DRAGONFLY AND DAMSELFLY VIEWING.
PHOTO CAMERON ECKERT.

it's a volunteer-supported (citizen scientist and biologist driven) atlas that provides a centralized source of "scientifically accurate information for use in conservation, education, and research."[54] It's a site where citizens and experts work together to document wildlife, and the result is a "living atlas" that is always evolving with new images, content, and interactive maps. If you would like to contribute your photos, read the E-Fauna page on Citizen Science Contributions. Before your dragonfly/damselfly walk at Cheam Lake, go to the E-Fauna homepage, and choose the dropdown menu for Insects, then under "O" choose the order Odonata (dragonflies). Scan through the images, names, and interactive maps of British Columbia's dragonflies to begin learning about the species that reside in southern BC. You can use this for any species in this book, not just insects, and, in fact, you can use its "sister websites," E-Flora (the plants of British Columbia) and Biodiversity of BC as well. Dragonflies and damselflies can easily be confused with other species, and there is so much more to learn about them, so you may want to consider a resource such as *Introducing the Dragonflies of British Columbia and the Yukon* by Robert A. Cannings, published by the Royal BC Museum.

Representative dragonflies and damselflies common to Cheam Lake Wetlands Park include:

SMALL/MEDIUM-SIZED DRAGONFLIES:

Blue Dasher (late May–August, some to mid-September)
Western Pondhawk (May–September)
Eight-spotted Skimmer (May–September)
Four-spotted Skimmer (mostly May–August)
Cherry-faced Meadowhawk (June–September)

1. BLUE DASHER (MALE AND FEMALE).
PHOTO MARION SHIKAZE.

2. WESTERN PONDHAWK (FEMALE).
PHOTO KATHRYN MILLIGAN.

3. EIGHT-SPOTTED SKIMMER (MALE).
PHOTO TERRY CARR.

4. FOUR-SPOTTED SKIMMER.
PHOTO TERRY CARR.

5. CHERRY-FACED MEADOWHAWK.
PHOTO COREY BUNNELL.

LARGE DRAGONFLIES (DARNERS):

Paddle-tailed Darner (mostly July–September)
Canada Darner (mostly July–August)
Common Green Darner (late April–early September, but mostly July and August)

DRAGONFLIES AND DAMSELFLIES ARE MAINLY A TROPICAL GROUP, AND THEREFORE, IN BC, YOU ARE MOST LIKELY TO FIND THEM IN AREAS WITH WARM WATER AND FLYING DURING THE MIDDAY WARMTH. THIS PADDLE-TAILED DARNER REALLY DOES HAVE A TAIL THAT LOOKS LIKE A PADDLE. PHOTO MARION SHIKAZE.

CANADA DARNERS ARE AMONG THE LARGEST AND FASTEST-FLYING DRAGONFLIES IN BC. *TRY TO GET A GOOD LOOK AT THEIR LARGE COMPOUND EYES ON THE TOP OF THEIR HEADS.* PHOTO ANTHEA FARR.

THE FACT THAT DRAGONFLIES, LIKE THIS MALE AND FEMALE PAIR OF COMMON GREEN DARNERS, SIT STILL FOR PERIODS OF TIME MAKES THEM A GOOD SUBJECT FOR PHOTOGRAPHY. MOST OF THE IMAGES IN THIS CHAPTER ARE FROM PEOPLE WHO HAVE TAKEN UP PHOTOGRAPHY AS A HOBBY. PHOTO MARION SHIKAZE.

COMMON DAMSELFLIES:

Northern Spreadwing (mostly July–early August)
Northern Bluet (May–September)
Pacific Forktail (April–September)

Bumblebees

Have you ever tried to identify one bumblebee from another? Bumblebees sort themselves out in terms of habitat, so those you see on flowers in your neighbourhood are likely different from those atop Grouse Mountain in the summer, or from those in Osoyoos. They love home gardens and roadside flowering plants, so you rarely have to go far to find them. Low- or high-elevation wildflower meadows are popular bumblebee habitats, as are raspberry patches and the dandelions, clover, or the goldenrod at your local park. When you see them, and if you wish to identify one from another, Syd Cannings suggests the following questions to consider:

- → Is the abdomen orange or yellow? If so, where is it orange and where is it yellow?
- → Is its face black or yellow?
- → Does the bee have packets of pollen on its legs?
- → Where are you (your location)?

Most of us understand that bees are important pollinators (though hummingbirds, bats, mice, other insects, and even some mammals act as pollinators too), and it's because of pollinators spreading pollen from plant to plant that we have enough food to survive. Life on Earth depends on plants and pollinators, and you can thank a pollinator for *every third bite* of food you eat! More pollinators, including bees, means more berries, fruits, vegetables, nuts, and even chocolate to satisfy your palate.[55] Bees are providing a service, known as an *ecosystem service* (for more information on this concept, see the Bats in Peachland chapter in this book), and we are the beneficiaries of that service. It turns out that many plants also benefit from this service. A closer look at plants and bumblebees will also help you see how each has shaped the other's evolution in dramatic ways. The markings on a foxglove flower, for instance, provide an ultraviolet "landing strip" for insects such as bees that can see that specific wavelength of light. Other flowers such as orchids have trap doors that only open for bees of a certain weight, slippery surfaces to ensure a bee slides through an area of pollen, or rough textures that allow bumblebees to get a grip!

Several BC communities, including the Municipality of Delta, have recognized the value of bees by committing to a designation from Bee City Canada intended to inspire communities to take action to protect pollinators. The key benefit of such programs is that they provide a starting point for conversations about the importance of biodiversity and how it is connected to growing healthy food and naturalizing our homes and community. They

DID YOU KNOW THAT POLLINATORS ARE RESPONSIBLE FOR EVERY THIRD BITE OF FOOD YOU EAT? BERRIES, FRUITS, VEGETABLES, NUTS, AND EVEN CHOCOLATE ALL EXIST BECAUSE OF THE WORK OF POLLINATORS. PHOTO ROB ALEXANDER.

also raise awareness of ways to reduce and eliminate chemicals that are harmful to pollinators, while providing resources to educate residents about habitat restoration. For example, we can ensure that insecticides, herbicides, and fungicides are used in our gardens and communities only when absolutely necessary. There is still so much to learn about bees, just as there is for other BC insects. Making our communities more beneficial for insects is a small step with big potential gains.

Representative bumblebees of Metro Vancouver and the Fraser Valley:

Orange-tailed (spring)
Sitka (spring-summer)
Black-tailed (spring-summer)
Tri-coloured (spring-summer)
Yellow-faced (late spring-summer)
Bright-yellow (late summer)

1. ORANGE-TAILED BUMBLEBEE ON OREGANO. PHOTO ROB ALEXANDER.
2. SITKA BUMBLEBEE (MALE). PHOTO ROB ALEXANDER.
3. BLACK-TAILED BUMBLEBEE. PHOTO TERRY CARR.
4. TRI-COLOURED BUMBLEBEE ON BLACKBERRY FLOWER. PHOTO ROB ALEXANDER.
5. YELLOW-FACED BUMBLEBEE. PHOTO TERRY CARR.
6. BRIGHT-YELLOW BUMBLEBEE. PHOTO ROB ALEXANDER.

> I love learning. The more you know, the more you learn, and the more you enjoy being outdoors. You see patterns and connections. It adds to your enjoyment of the whole ecosystem. When I travel, I see connections with what I learned at home and it opens a door to that new place. I can't imagine travelling somewhere and not asking questions. It's a way of placing yourself in the world. Inevitably it leads to a conservation ethic as you begin to be more empathetic to a species plight.
>
> —**SYDNEY CANNINGS,** naturalist, author, and scientist

Engaging and Taking Action

Do you have a lawn? Are you frustrated with crows ripping it up as they forage for European Chafer Beetles? Consider pulling out your grass and replacing it with insect-friendly plants. Since the loss of wetlands has had such a large impact on insects, you can make a difference by planting a

THE YELLOW-FACED BUMBLEBEE IS ONE OF NUMEROUS BRIGHT AND DISTINCT BUMBLEBEES. ***THIS YEAR CONSIDER LEARNING HOW TO DISTINGUISH ONE TYPE FROM ANOTHER. LOOK FOR THE DIFFERENCES IN THE HEAD, ABDOMEN, AND RUMP, AND IF YOU ARE OBSERVANT, YOU MAY EVEN SEE THEIR POLLEN BASKETS ON THEIR LEGS.*** PHOTO ROB ALEXANDER.

garden for butterflies, bees, and dragonflies, and even more so by building your own pond, as even a pond less than a metre across is large enough to attract decent-sized dragonflies. (But don't put in goldfish, as they eat the larvae!) Homeowners can create both summer and winter habitat for bumblebees by leaving leaf litter or rotting logs for bees to hibernate under, or by creating ground with hollows or burrows for nesting. Nests are used for egg laying, and typically colonies are not large (50–200 queen and worker bumblebees). A nest in your yard allows you to watch their annual cycle, which begins in late March to early April as they feed on flowers while gathering nectar and pollen. It continues as they build their underground colony and lay eggs (all the eggs are queens at this time), to the birth of the reproductive generation (males and new queens) of mid-August, and finally to the hibernation of the queens (ideally in your leaf litter). *The Nature of Things* (sans David Suzuki) right there on your property!

There is so much else that you can do for insects, from looking into local invasive species "plant pulls" (such as BroomBusters) to native species community plantings. Create a checklist of what's on your property, identify the flowers, shrubs, and trees around your home, then learn the connections between those and the birds, butterflies, and other insects that use them. Acquire a camera with a zoom lens to photograph the butterflies, dragonflies, or bumblebees you come across. Capture multiple images from various angles, note your location, attempt to identify them, and post your photos to iNaturalist.ca. If you are unsure what the species is, use the iNaturalist app or inaturalist.org to identify the insects that you see and to record their locations. If you wish to post further, consider posting to other online communities such as bumblebeewatch.org, or to E-Fauna as described earlier. Insect specialists will view your posting and likely provide an identification. Your contribution does make a difference to the understanding of a species range and habitats. In turn, this will help you make deeper connections, and if you share your learning with a wider circle you will be fertilizing the minds of others about these connections and the plight of insects in general.

Minimizing Your Carbon Footprint

Depending on your location, consider taking transit, walking, or cycling to a viewing location. When we visited Ross Lake campground in Skagit Valley Provincial Park, we noticed two BCAA Evo car share Prius hybrids, a reminder that even relatively remote roads like this are accessible to car-sharing programs. In Vancouver, the floral and water areas of Queen Elizabeth Park in Vancouver, for instance, harbour butterflies, bumblebees, and dragonflies in the summer. The park is easily accessible by alternate transport means.

Directions and Accessibility

Cheam Lake Wetlands (dragonflies) can be accessed via Highway 1 at the Bridal Falls Exit 138. Cross to the north side of the overpass. Turn left on Yale Road and east toward Popkum Road. Turn right on Popkum and left on Elgey Road. The park has a wheelchair-accessible viewing platform and boardwalk trail close to the parking lot.

Skagit Valley (butterflies) can be accessed via Highway 1 at Exit 168, Flood Hope Road, turning right onto Silver Skagit Road. Follow the Skagit Valley Provincial Park signs 60 km (approximately 1.5 hours' drive on gravel road) to the Ross Lake campground and be sure to watch for dragonflies while there too. Check the park website for accessibility options.

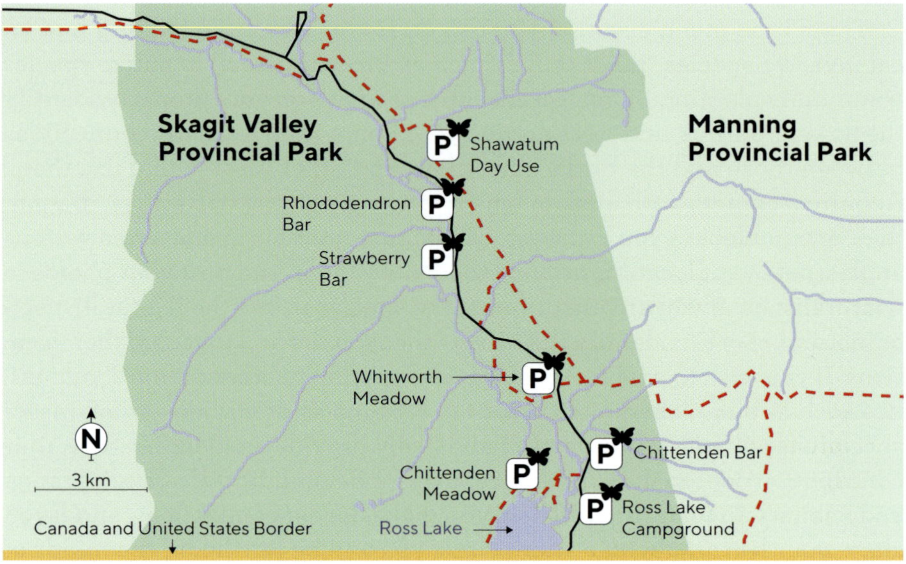

SEARCH FOR BUTTERFLIES ALONG THE CORRIDOR ROAD, AND ESPECIALLY IN THE PARKING AREAS, BUT BE CAREFUL TO PULL YOUR VEHICLE WELL OVER TO THE SIDE OF THE ROAD IF YOU STOP. SOURCE: BC PARKS, SKAGIT VALLEY PROVINCIAL PARK.

Other Locations to View Insects

1. UBC has the Spencer Entomological Collection housed at the Beaty Biodiversity Museum. It's a great starting place to get to know your insects and there are interpreters there who may provide further insight into the displays.

2. Queen Elizabeth Gardens in Vancouver attracts many butterflies to its floral display.

3. Butterfly World in Coombs or Butterfly Gardens in Victoria are both private businesses that offer close-up viewing of butterflies.

4. Minnekhada Regional Park in Coquitlam on the trail between the marshes is a great dragonfly spotting area in midsummer.

Additional Resources

1. If you want to consider planting a garden for butterflies, dragonflies, or bees, check out **BC Wild Heritage Plants** (Flora of the Pacific Northwest) or West Coast Seeds.

2. Consider checking out the **UBC Geography's** flora, fauna, and biodiversity sites. The sites include: E-Flora, E-Fauna, and Biodiversity of BC.

3. **Nature Chilliwack** has a Facebook photo page for Butterflies Host and Nectar Plants organized by Denis Knopp that is very useful.

4. Norbert Kondla, from the University of Calgary, is a well-respected specialist of butterflies and his website is packed with great content. You can check it out at **Monarchbutterflies.ca.**

5. ***Bumble Bees of the Western United States*** is also a very useful resource for bees in southern BC.

6. ***Life Histories of Cascadia Butterflies*** by David G. James and David Nunnallee presents the life histories of virtually all of the 158 butterfly species occurring in southern British Columbia. This is a good resource if you want to plant a garden for coastal native butterflies.

CLIMATE CHANGE AND INSECTS

Reasons for insect declines are multifaceted and include agriculture, pesticides, habitat loss, pollution, forest fires, and extreme weather events. Each of these are likely to interact with climate change. Add an extreme drought that reduces nectar sources for butterflies, or has a drying effect on ponds used by dragonflies, and significant declines can happen quickly. Climate change also can lead insect populations to be "out of sync" with their predators, such as birds, with potentially serious ramifications for the greater ecosystem.

NOTICE THE PATTERNS AND SCALY SKIN OF THESE WESTERN RATTLESNAKES THAT ARE AMASSED FOR WARMTH IN THIS PHOTO. PHOTO CHLOE HOWARTH.

Western Rattlesnakes & Great Basin Gopher Snakes

at the Nk'Mip Desert Cultural Centre, Osoyoos, South Okanagan

AT A GLANCE

WHAT: Rattlesnakes (*Crotalus oreganus*) & Gopher Snakes (*Pituophis catenifer deserticola*)
WHERE: Nk'Mip Desert Cultural Centre, Osoyoos
WHEN: Summer, but especially on World Snake Day, July 16
TRADITIONAL TERRITORY: Osoyoos Indian Band, Sylix Okanagan Nation Territory
ACCESS: Moderate

What is your attitude about snakes? Love them? Fearful? Apprehensive? Excited? I feel a visceral sensation regarding snakes, and I think it includes all of these emotions. Because there are seven different types of snake in the South Okanagan Similkameen region, I love knowing I don't have to go to Arizona or Nevada to see one in the wild. At the same time, I'd be apprehensive walking in the desert on my own, knowing a camouflaged rattlesnake might be sunning itself on a rock next to where I'm about to step. But I'd still be *excited* to know it's there! I'll call it "cautious enthusiasm." In this chapter I'd like to emphasize the caution we must all take around snakes, while at the same time encouraging you to visit the controlled setting of the Nk'Mip (pronounced in-kuh-meep) Desert Cultural Centre in Osoyoos. There, you can experience the excitement of seeing snakes in a wild, but safer, setting. The centre is a reasonably reliable location to see a desert snake on its trails, so if you place yourself there at the right time and alongside their expert staff and volunteers, your chances

A VOLUNTEER INTERPRETER-LED SNAKE WALK ON THE ANTELOPE BRUSH GRASSLAND TRAILS SMARTLY STARTS AT THE NK'MIP DESERT CULTURAL CENTRE. PHOTO NK'MIP DESERT CULTURAL CENTRE.

RESOUNDING CLICKS OR PINGS OF RADIO TELEMETRY HELP TO LOCATE SPECIALLY TAGGED SNAKES. THE TAGS ARE CRITICAL TO HELPING RESEARCHERS UNDERSTAND THESE REPTILES' MOVEMENTS. PHOTO CHLOE HOWARTH.

THE OSOYOOS INDIAN BAND'S NK'MIP DESERT CULTURAL CENTRE IS PART OF THE PROPERTY THAT ALSO INCLUDES AN INDIGENOUS-OWNED WINERY AND TOURISM ACCOMMODATION. PHOTO NK'MIP DESERT CULTURAL CENTRE.

of encountering some kind of snake are high. That said, please realize that rattlesnakes are extremely shy, and though they cannot hear, they feel the vibrations of our feet as we approach, so they do what they can to hide and stay out of our way. That's what makes a guided walk so helpful.

Since 2007 the Nk'Mip Desert Cultural Centre, located on the Osoyoos Indian Reserve and run by the Osoyoos Indian Band, has been the epicentre of much of the research into snakes in BC. Scientists, along with graduate and undergraduate students, use this landscape to continue the longest-running snake conservation project in British Columbia, as they examine the long-term impacts of development on species at risk. This research helps increase our understanding of how snakes live, the habitats they need to survive, and how we can manage both humans and the landscape in ways that support safe snake-human interactions. While at the centre, don't be surprised to see or learn about radio telemetry, wildlife cameras, and snake rookeries and hibernacula, plus PVC "refugia" chambers built to provide shelter from high temperatures to snakes in areas of poor-quality habitat and along snake exclusion fencing. Researchers focus on topics such as: snake reproduction; changes to "rattling defensive behaviour" in areas of high human presence; the success of road underpasses fed by funnel fences to decrease snake mortality on the roads; and other mitigation measures being used to counter the challenges created by human development. These research studies, along with the associated education around them, led to the centre receiving a national conservation award in 2018 from the Canadian Herpetological Society. It's critically important work because the snakes in this Okanagan "**pocket desert**," one of BC's **biodiversity hotspots**, face many challenges, and credit is due to the Osoyoos Indian Band for making their land a hotspot for research and education!

In late March, snakes leave their **hibernacula** (winter dens, plural of hibernaculum) in the higher elevation rock bluffs above Osoyoos Lake to spend their summers in their customary foraging and mating grounds below, often travelling up to 1 km each day – that's a lot for a snake. Their movements, however, lead them to roads and disturbed landscapes such as residential areas, agricultural lands, and recreational sites like golf courses and campgrounds, which in turn leads to encounters with another species – humans. When snakes are crossing a road and they sense a vehicle approaching, they tend to freeze, a strategy evolved to "hide" them from a threat in the surrounding landscape, but on a road this behaviour doesn't work so well and makes them highly vulnerable to being squished and killed by tires.[56] That means roads built close to snake feeding, mating, and den areas pose an especially significant threat. In the region of White Lake just west of Okanagan Falls, Stephanie Winton, a former graduate student from Thompson Rivers University, has quantified the impact of roadkill patterns on overall mortality and has provided the data required to support potential solutions, like funnel fencing and road underpasses (also called "eco-passages"). Note I just said "potential." We tend to see human-made structures such as underpasses as solutions, but that is only the case if they can make enough of a difference to drastically decrease the amount of roadkill. Research is currently being done to assess the success of underpasses, and Winton told me that "If you're a motorist and you see a dead animal, multiply that by three (as scavengers remove many dead snakes from the roads). That's the true magnitude of the problem!" Therefore, Stephanie's advice is "Don't forget to brake for snakes."[57]

This threat of roads is especially important, as the region's long, cold winters and short summers mean female rattlesnakes have low reproductive rates and can only reproduce every two to three years. Typically, they have small litters of only two to eight young each time, and they just can't sustain the high mortality that vehicles present. Replenishing numbers can be a long process, especially when roads are generally added to a landscape, not removed. Add to that the additional threats of direct persecution by humans (note: it's illegal to harm, kill, or capture a rattlesnake in BC) and the continuing loss of habitat, both of which have contributed to keeping rattlesnake numbers from replenishing significantly so far.[58]

CULVERTS ARE INSTALLED TO ALLOW SNAKES TO PASS UNDER ROADS. ROADKILL IS A MAJOR THREAT TO SOUTH OKANAGAN SNAKES. PHOTO JADE SPRUYT.

FUNNEL FENCES HAVE BEEN INSTALLED ALONG ROADSIDES TO LEAD SNAKES TOWARD CULVERTS THAT PASS UNDER THE ROADS. THESE ARE NOW COMMON IN AREAS OF HIGH SNAKE MIGRATION. PHOTO JADE SPRUYT.

NOTICE THE PATTERNS AND SCALY SKIN OF THESE WESTERN RATTLESNAKES THAT ARE AMASSED FOR WARMTH IN THIS PHOTO. *HOW MANY INDIVIDUAL SNAKES CAN YOU COUNT?* PHOTO CHLOE HOWARTH.

The permanent loss of ecosystems, like the Antelope Brush shrub-steppe and grasslands, which are continually being transitioned to orchards, vineyards, and housing, ensure that the struggle to increase snake numbers will continue to be a challenge. To experience the Antelope Brush shrub-steppe ecosystem yourself, walk the trails of Nk'Mip and use the Seek App on your phone to learn more about the plants in this ecosystem. Snakes play a valuable and complex role in the shrub-steppe ecosystem, both as predators (of rodents) and prey (of hawks and eagles) and help maintain a balance as part of a wider food web of relationships between species. To experience snakes as part of the biodiversity of the shrub-steppe ecosystem, my suggestion is to be at Nk'Mip for World Snake Day on July 16, as there will be live exhibits, presentations by interpreters, and walking tours with researchers who will impart a deeper understanding of this ecosystem and raise awareness of the role of snakes within it. As you know, landscapes change with the seasons, so although I suggest a visit on World Snake Day, there are many reasons to visit at other times. The seasonal wildflower colours of Mariposa Lily, phlox, larkspur, and the bright yellow blossoms of the Antelope Brush support a diversity of butterflies and birds that are unique to this region, and they are reasons in themselves to visit Nk'Mip, while also being good times for snake viewing.

There are also many options for walks in the wider grasslands of the South Okanagan Similkameen beyond Nk'Mip, and if you do go out without a guide, consider wearing ankle-high boots and staying on trails that allow you to see your feet and the area around them. Additionally, don't reach your hands into any space you can't see into, as snakes like to hide under objects and in dark spaces. If you do hear a snake rattle, stop immediately and look around you, as there may also be other snakes in the vicinity. Then move carefully away, giving the snake a space of at least 5 m. Before you leave home, consider learning to differentiate between the venomous Western Rattlesnake and the non-venomous Great Basin Gopher Snake, although the latter does have a convincing rattlesnake imitation when approached. With a little effort, however, they are easy to tell apart.

You could also start your learning at bcreptilesandamphibians.ca and take their snake quiz! While there, you may learn there are approximately 300 adult Western Rattlesnakes in this region, with a density of about 58 adult rattlesnakes per square kilometre. Although adult survival from year

THE TONGUE OF A SNAKE IS USED TO GAIN AN UNDERSTANDING OF ITS ENVIRONMENT. FLICKING THE TONGUE ALLOWS THEM TO DETECT ODOUR-CAUSING PARTICLES OR PHEROMONES IN THE AIR, MUCH LIKE OUR NOSES DO. PHOTO CHLOE HOWARTH.

THE HUNTING STRATEGY OF THE GREAT BASIN GOPHER SNAKE IS SLIGHTLY MORE ACTIVE THAN THAT OF A WESTERN RATTLESNAKE, AS THE LATTER USES FAST-KILLING VENOM.

to year is around 85 per cent, snakes that live closer to human development are often in poorer condition. Like many other species outlined in this book (such as hummingbirds that return to your feeder each year), rattlesnakes have high site fidelity, meaning they return to the same places year after year. Hibernacula, where rattlesnakes overwinter, are a good example of site fidelity. Females give birth every two to three years and may also return to their rookeries, the locations where they gave birth to their live young previously. When humans decide to develop on land, the potential for irreparably altering a high-fidelity site or blocking access to it is significant. Obviously, this can be devastating for the snakes as humans always win these battles in the short term. In order to address this problem, Dr. Christine Bishop, a research scientist in the wildlife research division of Environment and Climate Change Canada, is testing artificial snake dens at Nk'Mip to determine if they could provide additional shelter for snakes in areas that experience more human-caused stress, such as those with barriers to movement. But make no mistake, nothing compares to wild habitat.[59]

Although there are seven snake species in the South Okanagan, I've chosen only two of the most common to profile in this chapter. The first is the Western Rattlesnake, which has a triangular-shaped head, a tan to pale green stout body, oval spots on its back, and grows to between 0.6 and 1.2 m as an adult. In contrast, the Great Basin Gopher Snake (also called a bull snake) grows up to 1.8 m in length, with a sleek head and dark line (like a mask) running across its eye, and a creamy yellow to tan slim body with dark-coloured rectangular patches. Food for both of these "**ectothermic creatures**" (animals that rely on the environment to maintain their body temperature) consists of small mammals, such as deer mice, lizards, and birds and their eggs. The gopher snake has a slightly more active hunting strategy due to its lack of venom, while the rattlesnake is more passive, waiting for prey to pass by. A larger snake needs a larger meal, meaning a full-size adult gopher snake can even have a marmot on its menu! Both species hibernate and, interestingly, multiple snake species appear to make a peace agreement to keep fighting out of their dens, allowing them to share space in winter hibernacula. One significant difference between these snakes is that rattlesnakes bear live young, usually around the month of August, while gopher snakes lay eggs around the end of June with young hatching in late summer. Newborn rattlesnakes are fully venomous, but they lack a rattle, as the rattle is created by skin shedding. With each skin shed, which can happen one to three times per season, a new piece of hardened skin (keratin, like our fingernails) is added to the button on their tails, creating a new segment that sits loosely inside the next, creating the rattle. Older snakes can have many rattles, but they do break off (just like our fingernails can crack)

THE ANTELOPE BRUSH ECOSYSTEM DOMINATES THE AREA AROUND OSOYOOS. BECAUSE IT SUPPORTS WESTERN RATTLESNAKES, *BE CAREFUL IF WALKING OFF TRAIL IN THIS LANDSCAPE. OR BETTER YET, STAY ON THE TRAIL!* PHOTO CHLOE HOWARTH.

over time too. Both species require three critical habitats: hibernacula, birthing or egg laying sites, and summer foraging and mating grounds.[60]

To ensure conservation of these critical habitats, the British Columbia Ministry of Environment and Climate Change Strategy has created a system across the province, for habitat protection beyond our provincial parks and **ecological reserves**, called Wildlife Habitat Areas, or "WHAs" for short. These are critical habitats in which activities are managed to limit impact on identified wildlife species, with the ultimate purpose of conserving habitat that is potentially most limiting to that particular species, like rattlesnakes. In the Okanagan region alone, there are 433 WHAs, and eight of them are in the Antelope Brush shrub-steppe community discussed in this chapter. The Osoyoos Desert Centre is one of those eight, and this means it has limitations such as access (road building), as well as limitations on pesticide use, livestock management, and recreation. These WHAs provide protection in the South Okanagan for species that range from

Coastal Tailed Frogs to Grizzly Bears, and from Lewis's Woodpeckers to Half-moon Hairstreak Butterflies.[61] The South Okanagan is home to 30 per cent of BC's red-listed species and 46 per cent of its blue-listed ones, which include both snakes discussed in this chapter. It's not a given that species that are with us today will be with us tomorrow. Others that used to call the South Okanagan home, but have been extirpated (which means they don't exist in the region anymore), include the White-tailed Jackrabbit, Burrowing Owl, Northern Leopard Frog, Sharp-tailed Grouse, and Sage Grouse.[62] More education, awareness, and support for conservation policy is critical for the species that remain, and snakes of the South Okanagan are among those that will require actively managed Wildlife Habitat Areas to survive.

> There is a great connection between species that can be considered while at the Centre. The Great Basin Pocket Mouse lives in this landscape. We also have Antelope Brush, a protected plant species. These species rely on each other – the pocket mice eat Antelope Brush seeds and cache the seeds down in their borrows. There, any seeds that don't get eaten germinate into new shrubs. The Antelope Brush feeds the pocket mouse, and in return, the pocket mouse disperses the Antelope Brush seeds, ensuring its survival. This is also important because small mammals like the pocket mouse are the main prey species of the Western Rattlesnake.
>
> —**CHLOE HOWARTH,** master's student and rattlesnake researcher

Engaging and Taking Action

Taking action for snakes may be less "hands on" than for other species outlined in this book, but there are always opportunities. Simply educating yourself is an important action to take, allowing you to understand at a deeper level how snakes are part of a wider ecosystem, while dispelling myths about them. As outlined above, Nk'Mip Desert Cultural Centre offers an array of walks, presentations, and events; they also have an Adopt-a-Snake program that provides a way for you to learn about snakes, while also making a donation that supports their well-being. You can also use the app iNaturalist, widely used by scientists, to record your sightings as a citizen scientist, which helps in ongoing science research. If you encounter a rattlesnake and you would like to add this information to the province's database for conservation research, please use the form located on the provincial environment website at the Submit Wildlife / Plant Data and Information page. Still another option is to consider a wildlife viewing trip to the South Okanagan that is arranged around the

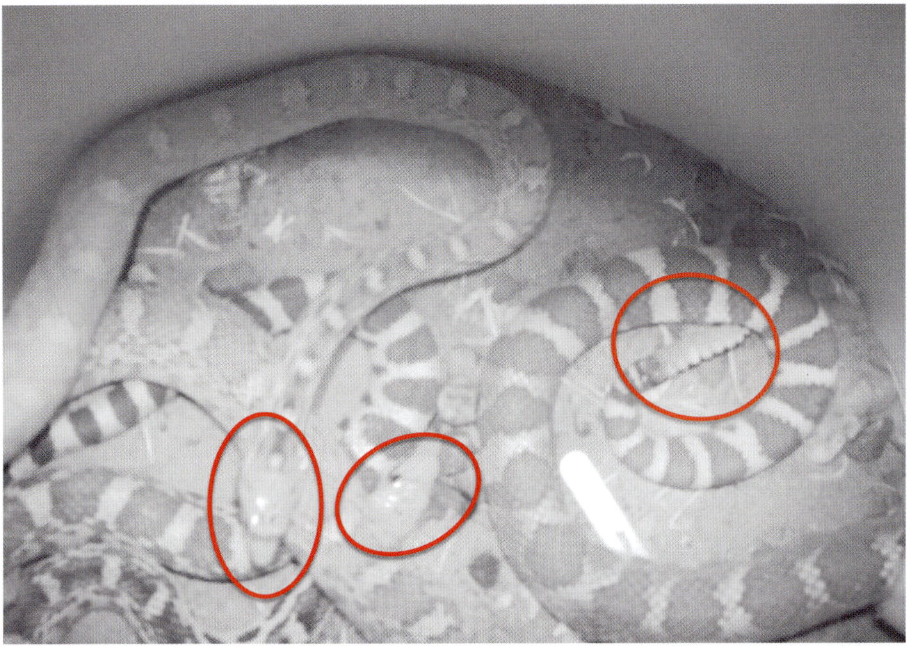

PVC BUCKETS, USED AS "REFUGIA" CHAMBERS, PROVIDE SHELTER FROM HIGH TEMPERATURES FOR SNAKES IN AREAS OF POOR-QUALITY HABITAT AND ALONG SNAKE-EXCLUSION FENCING AREAS. THIS ONE HAS A WILDLIFE CAMERA. RED CIRCLES NOTE THE LOCATION OF THE HEADS OF TWO GOPHER SNAKES, AS WELL AS A RATTLESNAKE RATTLE. PHOTO CHLOE HOWARTH.

projects run by the Okanagan Stewardship Society and/or the Okanagan Similkameen Conservation Alliance, which also runs the Meadowlark Nature Festival (see the chapter on the Meadowlark Nature Festival). And, of course, taking action means sharing your knowledge with others, including repeating Stephanie Winton's slogan, "Don't forget to brake for snakes," which should be passed on to all Okanagan-bound travellers!

I asked Dr. Karl Larsen, professor in the Department of Natural Resource Sciences at Thompson Rivers University, how he thought people could take action regarding snakes in the Okanagan. Without hesitation he said people need to help stop the spread of invasive species. It was an answer I didn't expect regarding snakes, but on reflection it made sense considering the South Okanagan ecosystem. **Invasive species** of any kind, plant or animal, can disrupt the functioning of ecosystems and natural food webs for endangered ecosystems like that found in the South Okanagan. This can be especially problematic when the balance of an ecosystem is already so delicate and impacted by other issues such as development. Ultimately, invasive species can have **cascading effects** throughout the rest of the food web. His point was that even caring people unintentionally exacerbate invasive species problems, and gave an example of well-meaning people who feed birds, and thus unintentionally feed squirrels. In the South Okanagan there are two invasive squirrels (the Eastern Grey and the Eastern Fox) that are having larger litters and spreading more rapidly because of the extra food supplied at bird feeding stations. It's yet *another reminder that our actions matter.* If you take this to heart, then *Lim' Limpt*. That's *thank you* in the language of the Syilx People of the Okanagan Nation.

Minimizing Your Carbon Footprint

At 400 km from Vancouver, Osoyoos is a significant distance to go to view wild snakes, but it's worth the trek, particularly if you can travel using a carbon-light option. Mountain Man Mike's Bus Service is one of those options, taking approximately 5½ hours to travel between the Braid SkyTrain Station in New Westminster and Princeton, and onto Osoyoos, twice a week. Check out their schedule online. Osoyoos itself is a very walkable town, and Nk'Mip is only a 4 km, 45-minute walk from downtown.

Directions and Accessibility

Nk'Mip Desert Cultural Centre is on the east side of Osoyoos. It has a wheelchair ramp from the parking lot. The building is fully wheelchair accessible and their outdoor exhibits have accessible gravel crush pathways, as does 1 km of nature trail going to the village displays. Some areas of rock crush trail, such as the steeper upper trail, may be difficult for motorized wheelchairs or scooters. Wheelchair-accessible and non-accessible trails are clearly marked on the outdoor trail map. In addition, seniors or special-needs guests can ask if staff are available to conduct a tour by golf cart.

Other Locations to View Snakes

1. The Osoyoos Desert Centre is run by the non-profit Osoyoos Desert Society with a mandate to conserve, restore, and educate. They strive to generate public knowledge, respect, and active concern for the South Okanagan's fragile and endangered ecosystems. At the centre you can take a guided or self-guided tour along a 1.5 km elevated wooden boardwalk and visit an interpretive facility.

2. Okanagan Falls Provincial Park and the White Lake Grasslands Protected Area near Okanagan Falls are both good places for seeing snakes in the wild, but please remember to review the safety precautions outlined in this chapter before you go.

3. The Rattlesnake Bluffs Conservation Area in the Lac du Bois Grasslands near Kamloops is also a good location to see rattlesnakes (see page 332). If you decide to go to that area, also be sure to read the chapter about California Bighorn Sheep in that area, for a two-for-one viewing experience as outlined in another chapter of this book.

4. Snakes of the South Okanagan and Similkameen include: Night Snake, Rubber Boa, Racer, Western Rattlesnake, Great Basin Gopher Snake, and the Common and Terrestrial Garter Snakes.

5. Often children who are interested in snakes, are also fascinated by other reptiles and amphibians. Be sure to check out the content in this book on the Western Painted Turtle in the "Other Area Interests" section of the Chum Salmon chapter. Also, in March through April, at North Vancouver's Rice Lake in the Lower Seymour Conservation Reserve, you should be able to hear Pacific Tree Frogs and see Rough-skinned Newts.

THE PACIFIC TREE FROG IS QUITE COMMON AROUND URBAN AREAS IN SOUTHERN BC'S WOODLANDS, MEADOWS, AND PASTURES. ***LISTEN FOR ITS CLASSIC "RIB-BIT" IN MARCH AND APRIL, AND ITS "C-R-R-ICK" CALL THE REST OF THE YEAR.*** PHOTO ROB ALEXANDER.

ROUGH-SKINNED NEWTS ARE POISONOUS IF INGESTED. IT'S BEST NOT TO HANDLE THEM, BUT IF YOU SEE THEM TRYING TO CROSS THE ROAD, PLEASE PICK THEM UP AND MOVE THEM TO THE SIDE IN THE DIRECTION THEY ARE HEADED. ***LOOK FOR THEM IN THE SPRING AND FALL GOING TO OR FROM WATER AREAS.*** PHOTO SHELLEY FRICK.

Other Area Interests

1. 2¼ hours northwest and 184 km away from Osoyoos is Kentucky-Alleyne Provincial Park. Although this may be too far to be considered an "Other Area Interest" for Osoyoos, I wanted to lead thoughtful developing naturalists to this amphibian viewing site for tiny Western Toad "toadlets," but only if you're willing to be super careful! And you may even come across some snakes here, too. So far, the area hasn't been protected enough to ensure the toadlets aren't trampled by the uninformed public, so it is critical that potential wildlife viewers protect this fragile, yellow-listed Western Toad breeding site. Kentucky-Alleyne park has a reservable campground, and if you reserve for the first week of August, and if it happens to have been raining, you may be there to witness thousands of fingernail-sized toadlets moving from the west pond, near the isthmus of the pond and Kentucky Lake, as they migrate toward the forest. Watch for the low fencing set up by the Nicola Valley Naturalists to funnel the toadlets toward a tunnel underpass, which is intended to keep the tiny creatures from crossing the BC Park campsite road. If toadlets are evident, please stay out of the tall grass and stick to the paths, then walk slowly, waving your foot at each step to ensure there is nothing moving underneath.

2. Sw̓iw̓s (Haynes Point) Provincial Park is among the most popular campgrounds in the South Okanagan.

Additional Resources

1. ***Pocket Desert – Confessions of a Snake Killer*** is a National Film Board of Canada production that adds a good personal story to the content of this chapter.

2. Thompson Rivers University has a **Reptiles of BC** site that is excellent. Check it out at bcreptilesandamphibians.ca.

3. ***Spotlight on Species*** is a worthwhile online PDF document providing great content on Snakes of the South Okanagan Similkameen.

4. **Biodiversity Research Centre at UBC** – The goal is to understand and conserve the diversity of life on Earth through research, education, and outreach.

5. ***"Help Stop the Spread of Invasives"*** – Do you know what's growing in your garden? This is a Metro Vancouver Parks Brochure written in collaboration with The Greater Vancouver Invasive Plant Council.

CLIMATE CHANGE AND SNAKES

Snakes are ectothermic animals, relying heavily on their environments and the habitats they select in order to maintain their optimum body temperatures. Warming effects of climate change in the Okanagan may alter the way snakes are able to use their habitat and might even lead to heat-stressed snakes from farther south migrating northward, upsetting the ecosystem balance. In addition, climate change is likely to impact the availability of prey species snakes rely on, with cascading consequences.

YELLOW-BELLIED MARMOTS ARE THE MOST COMMON MARMOTS FOUND IN THE SUBALPINE AREAS OF BC. LISTEN FOR THEIR HIGH-PITCHED WHISTLE CALL AND WATCH FOR BEHAVIOURS THAT CONNECT THEM TO THEIR WIDER COLONY. PHOTO JAMES ST. JOHN, FLICKR, CC BY 2.0.

Small Mammals: Squirrels, Marmots, & American Pikas

at E.C. Manning Provincial Park

AT A GLANCE

WHAT: Columbian Ground Squirrel (*Urocitellus columbianus*), Cascade Golden-mantled Ground Squirrel (*Spermophilus saturatus*), Yellow-bellied Marmot (*Marmota flaviventris*), American Pika (*Ochotona princeps*). Also look for Red & Douglas Squirrels, and Yellow-pine Chipmunk

WHERE: Manning Park

WHEN: July, in alpine wildflower season

TRADITIONAL TERRITORY: S'ólh Téméxw (Stó:lō), Syilx tmix^{w} (Okanagan), Nłeʔkepmx Tmíx^{w} (Nlaka'pamux)

ACCESS: Easy to moderate

Growing up, how did your family spend summer holidays? In raising our family, we explored a variety of new geographies and ecosystems each summer, and therefore my children didn't have much experience of returning to the same location year after year. E.C. Manning Provincial Park, however, which lies in the heart of the Cascade Mountains but is just under three hours from Vancouver, is one of those locations that draws people into annual repeat visits, sometimes even spanning multiple generations! Many come to the trails, to experience a different ecosystem from the coast, or for the lake recreation, camping, and wildlife. In my recent research into the park, I discovered something interesting. The Columbian Ground Squirrels, the creatures that frequent the grass field near the picnic area outside the resort, provide a special attraction, which "totally draws people to the park."[63] That's a quote from Jo Hughes, the head naturalist at Manning Park, and she opened my mind to how many people

IDENTIFIED BY STRIPES THAT EXTEND THROUGH ITS FACE, THE YELLOW-PINE CHIPMUNK IS PROBABLY THE MOST EASILY RECOGNIZED SMALL MAMMAL IN MANNING PARK. AS WITH OTHER SPECIES IN THE PARK, PLEASE DON'T FEED CHIPMUNKS, EVEN IF THEY COME TO YOU LOOKING FOR A HANDOUT. PHOTO MARION SHIKAZE.

watch squirrel antics as they pop up and down burrows, interact with each other, and also with us. The act of watching is a "communing-with-nature" experience, and it takes place right at the edge of Highway 3! This type of experience isn't limited to the grass field by the resort, however. It extends to the Lightning Lake day use area, with more ground squirrel colonies; to the Cascade Lookout with both Yellow-pine Chipmunks and the Cascade Golden-mantled Ground Squirrel; and to Blackwall Peak with the Yellow-bellied Marmot. Manning Park is a small mammal paradise, and certainly one of the easiest locations to see an array of species with little effort. But we need to talk. There is one action that doesn't go well with a "communing-with-nature experience," and that's *feeding of these animals*.

First, under the **BC Wildlife Act**, it is *illegal to feed wildlife* in a BC provincial park. Next, respecting and protecting wildlife ultimately should be a wildlife viewer's goal. BC Parks has two educational brochures on this topic called "*Signs You Are Too Close to Wildlife*" and "*Protecting Wildlife.*" Even if it were okay to draw animals closer to humans and their roads, infrastructure, and array of human behaviours, the fact is our snacks and lunches don't provide the nutrients the animals need to make it through the winter. Furthermore, feeding animals processed Cheezies, bread, or chips in summer fools their bodies into thinking they have met their energy stores for the winter too early, sending them into hibernation by mid-August. If you arrive at the park later in the summer, the Columbian Ground Squirrels near the resort are nowhere to be seen, as they have already gone down into their burrows for hibernation, the result being that they sleep while nature's food is still available, with a long winter ahead and Cheetos in their bellies. Many will fail to successfully bear young, and others won't survive until spring, all because they were fooled into "going to bed" too early.

Choosing whether or not to feed wildlife is a complex decision, so take the time to inform yourself about best practices (for the animals). As we all know, many bird lovers put out seeds and millet for songbirds, nectar for hummingbirds, or suet for woodpeckers, and in doing so they feel a real connection with wildlife, sometimes seminal connections that lead to a lifelong love or career. Personally, I take great delight in watching an Anna's Hummingbird sitting at my office window feeder, and I will never forget what a chickadee or Canada Jay (Whiskey Jack) feels like sitting on my hand as it eats the offered sunflower seeds. And more than ever, it's necessary to bridge the growing gap between children and nature, as devices take over so much of their time, and we can't ignore the bond that is created when an animal interacts with a child. Therefore, thoughtful consideration and critical thinking are necessary to discern which of our actions might be appropriate. A good strategy regarding feeding birds would be to talk with the experts who run the Wild Birds Unlimited stores or the Backyard Bird Centres. We should never have an intention to tame a wild species. Animals have a natural tendency to be uneasy around us and that's critical for their survival. And remember that our food waste, such as apple cores indiscriminately thrown out a car window or orange peels left at a viewpoint, often bring wildlife onto roads and close to people, as they become bolder, tamer, and seek more treats. If we put their *wild lives* at risk, it leads to diminishing returns for them! Remember most of Manning Park's small mammals have a hibernation to prepare for, and if they don't have the right nutrients and the right fats, they may not see next spring.

The hibernating burrows of the Columbian Ground Squirrel are built below the frost line and their nests are lined with grass and closed off with dirt at the entrance. As with other hibernating species, their heart rates, respiration rates, and body temperatures all drop significantly during hibernation, allowing their bodies to expend less energy through the long, cold months.[64] Survival in hibernation comes down to what condition they are in when they go down for the winter. Let's consider ground squirrels. If and when they emerge from their burrows in late March or early April, the ground squirrels are lightweights, and they must begin to build up their body fat immediately.

A DOUGLAS SQUIRREL'S PAWS OFFER THE DEXTERITY REQUIRED TO MANOEUVRE THE CONES AS THEY EAT THE SEEDS WITHIN THE BRACTS. PHOTO MICHAEL SCHMIDT PHOTOGRAPHY.

Those that are mothers are at a disadvantage as they must also feed and care for their new litters, which continue to nurse for about a month. Most of Manning Park's small mammals only spend a few months above ground, and after emerging they spend their time foraging for seeds, roots, bulbs, fungi, leaves, grasses, insects, and berries.

When you arrive at Manning Park, take time to observe what the ground squirrels are feeding on. At first, you may not notice that individual members of the greater colonies can be distinguished by their physical characteristics, but as you spend time observing you'll notice unique mixtures of grey, black, brown, and white hairs on their head, or perhaps a rusty patch above their nose. If you watch long enough, you'll observe their upright body stance as they eat with their front paws, and their darting movements as they speed from one location to the next. It's very entertaining! If you're lucky, a predator such as a hawk or eagle will pass overhead, and you'll hear their loud warning calls. The same may apply if there's an off-leash dog in the picnic area (please keep dogs on a short leash).[65] The truth is that this ground squirrel that we love to watch is a relative newcomer to the park and is in fact considered an *invasive species*. It arrived from the BC interior along the road corridor after Highway 3 was built in 1949, has expanded its range up the road to Blackwall Peak, and is currently moving west toward

Hope and the Fraser Valley. The Columbian Ground Squirrel, however, is just one of over 60 species of mammals that reside in Manning Park, so let's look at some of the others that thrive in this protected area.

In the mid-1990s the lower parking area at Blackwall Peak had several large boulders that supported a family or two of Hoary Marmots. These large (4.5 kg, 0.75 m tall) members of the squirrel family are what the Whistler resort was named after, referring to their shrill, whistling call. I learned from Kelly Pearce of Hope Mountain Centre that Hoary Marmots were last seen at Blackwell Peak in 2012, possibly due to a combination of the processed junk food thrown to them from people in their parked cars, coupled with an especially harsh winter. As Kelly said, "It's often people who make it difficult for a species to survive."[66] Although it's unlikely you will see the Hoary Marmot in Manning Park's front country, you will probably see

POPPING UP OR DIVING DOWN THEIR BURROW ENTRANCES, COLUMBIAN GROUND SQUIRRELS ARE OFTEN THE FIRST WILDLIFE THAT TRAVELLERS OBSERVE AS THEY PULL INTO THE PARKING LOT ADJACENT TO THE GRASSY AREA AT THE MANNING PARK RESORT.
PHOTO ROB ALEXANDER.

A TRIP UP TO BLACKWALL PEAK PROVIDES ACCESS TO TRAILS IN THE SUBALPINE MEADOWS, FLUSH IN SUMMER WITH FLOWERING PLANTS SUCH AS LUPINE, PAINTBRUSH, VALERIAN, ARNICA, AND HELLEBORE. PHOTO CHRISTINE GAIO.

Yellow-bellied Marmots when walking a 4 km loop trail (between the lower and upper parking areas, Paintbrush and Heather trails) to the trail junction during the "kaleidoscope of colour" wildflower season in mid-July.

Marmots, American Pika, and other subalpine species have evolved strategies for coping with the seasons over thousands of years, and though historical seasonal patterns can vary, they have generally been predictable. Short, mild summers produce bursts of vegetation and food, while long winters have shaped animals to hibernate and depend on fat reserves, or, in the case of many birds and some mammals, to migrate. Today, species

experience seasons that are less dependable than those their ancestors evolved in, and so their ability to survive is governed less by adaptations evolved over millennia and more by current capricious weather. Yellow-bellied Marmots, for example, are experiencing longer, warmer, and drier summers, which change foraging conditions, as well as warmer, shorter winters with less snowfall, which alters snow's insulating properties. In fact, our changing climate can result in beneficial conditions in one season, and difficult conditions in another, and this has potentially devastating consequences, especially for those eking out an existence in an already challenging alpine environment.[67]

Six kilometres up the Blackwall Peak road is Cascade Lookout, a reliable place to see the Cascade Golden-mantled Ground Squirrel. It looks like the Yellow-pine Chipmunk, also seen at the lookout, but is three times the size and lacks the facial stripes of its similar-looking cousin. Manning Park is the northern range limit of many species, and that is also the case for the Cascade Golden-mantled Ground Squirrel, which exists throughout the North Cascades but isn't found much farther north of this lookout.

The final two small mammals to consider on a trip to Manning Park may be more difficult to spot, but the American Pika and the Red Squirrel are both still abundant. If you are from the southwest coast and know the Douglas Squirrel, you'll notice that while the Douglas Squirrel has a rust- to cream-coloured belly, the Red Squirrel's belly is light grey to white. Red Squirrels are fairly easy to find in the campgrounds and on many trails, including the loop around Lightning Lake. Also remember to watch for their seed middens, discarded piles of half or fully eaten conifer cones, scattered throughout the forest. Pikas are often seen at the Cascade Lookout or on the rocks as you drive between the lookout and Blackwall Peak (keep your car window open to listen for their loud PEEP!). If you decide to walk the wildflower trails on Blackwall Peak, first listen for the American Pika, in rocky areas, then stop, wait, and scan the area you heard it. Within 60 seconds it is likely you will see it run! This also applies to the trail from Lightning to Flash lakes, where you may see Pika as you pass the two scree

OFTEN SEEN AT THE CASCADE LOOKOUT, THE GOLDEN-MANTLED GROUND SQUIRREL IS AT THE NORTHERN EDGE OF ITS RANGE HERE IN MANNING PARK. NOT TO BE CONFUSED WITH A CHIPMUNK, THIS GROUND SQUIRREL HAS A GOLDEN "CAPE," WHICH ACCOUNTS FOR ITS NAME. PHOTO FRANCO FOLINI, FLICKR, CC BY-SA 2.0.

WHILE IN THE PARK, CONSIDER LEARNING HOW TO USE iNATURALIST TO HELP DOCUMENT, IDENTIFY, AND CELEBRATE THE BIODIVERSITY OF SPECIES WHILE SHARING WITH A WIDER COMMUNITY OF OBSERVERS LIKE YOURSELF. **IT'S A GREAT CITIZEN-SCIENCE TOOL!** SOURCE: INATURALIST.ORG.

iNaturalist

slopes that meet up with the trail. Use the BC Park iNaturalist page to see the locations of the latest Pika sightings in the park. Pika, which are members of the rabbit family, are especially fun to view if they happen to be gathering grasses for the winter. They don't hibernate but instead use the summer to create strategically placed hay piles, in which they create tunnels to travel through once the snow arrives. They spend their winter awake under the snowpack, moving between their haystack feeding stations.

Beyond small mammals there are other species to observe in Manning Park, and I'd suggest seeking out information about the unique relationship between the Whitebark Pine (tree) and the Clark's Nutcracker (bird) and to understand them. It's a fascinating story. The visitor centre or one of the park naturalists can help with that! The Canada Jay and Common Raven should be easy to find (again because people feed them), and the hummingbirds, which are moving inland and north from the coast and the south, should be making their presence known in early summer as they buzz and click around the resort feeders.

IT'S ALWAYS A GOOD DAY WHEN ONE SEES AN AMERICAN PIKA ON AN ALPINE HIKE. THESE MEMBERS OF THE RABBIT FAMILY ARE OFTEN FOUND DASHING THROUGH BOULDER FIELDS CARRYING GRASSES TO STORE AWAY FOR WINTER. AS THEY DON'T HIBERNATE, THEY SPEND THE SUMMER CREATING FOOD CACHES TO GET THEM THROUGH THE WINTER.
PHOTO ERIC GROSS.

MANNING PARK HAS BOTH THE DOUGLAS SQUIRREL, SEEN ON THE BC COAST AND IN THIS IMAGE, AND THE RED SQUIRREL OF THE INTERIOR. *SEE IF YOU CAN DISCERN DIFFERENCES BETWEEN THESE TWO SIMILAR ANIMALS.* PHOTO ROB ALEXANDER.

HOPE MOUNTAIN CENTRE FOR OUTDOOR LEARNING AND MANNING PARK BOTH OFFER EXPERIENTIAL INTERPRETIVE PROGRAMS FOR CHILDREN, FAMILIES, AND ADULTS. *CHECK OUT WHAT OPTIONS ARE AVAILABLE DURING YOUR VISIT TO MANNING PARK, BY STARTING AT THE VISITOR CENTRE.* PHOTO KELLY PEARCE.

> I love meeting families that keep coming back to the park year after year after year. I've had kids who came to my program years ago, that now are parents and bring their own children. There are solid connections being made here and they span generations!
>
> —**JO HUGHES**, head of Manning Park's Naturalist Program

Engaging and Taking Action

Join the 10,000 or so people each year attending a nature program at Manning Park, because even if you know nature well, the naturalists there are likely to show you something new, or show you something familiar but in a new way. BC Parks also has a system for the public to volunteer in our provincial parks. Check out bcparks.ca/volunteers/.

Review the offerings from the Hope Mountain Centre for outdoor learning. Their mission is to provide outdoor education and ecological awareness opportunities in the watersheds of southwestern BC. Their programs range from winter snowshoeing and avalanche safety to backcountry cooking and a **Bird Blitz** at Manning Park. This non-profit, charitable organization also leads many wider conservation initiatives, including a Coast to Cascades Grizzly Bear Initiative, water testing, a BC Rivers Day event in September, and a Leave No Trace outdoor ethics instructor training program. The centre was also the main driver behind the historic Hudson's Bay Company Heritage Trail, a 74 km backpacking trail between Hope and Tulameen, following the same route taken by the HBC horse brigades of the mid-19th century.

Minimizing Your Carbon Footprint

Are you a hardcore cyclist? Manning Park is 217 km from, and 1300 m of elevation above, Vancouver (with an additional 600 m of elevation to Blackwall Peak). Some serious spring training and effort could lead to a summer "century" ride for a unique wildlife viewing experience. A bus is also available, and you can check the Manning Park Resort website for the most current information on that. If you do bus, see if you can bring along a bike, as the park and its trails are spread out but very cycle friendly.

Directions and Accessibility

Take Highway 1 to Hope and Highway 3 to Manning Park.

Manning Park has various accessible options around the resort, Lightning Lake, and Blackwall Peak. Contact the Manning Park Resort for more information.

Other Locations to View Small Mammals

There are many locations in southern BC but two other locations specifically mentioned in this book include the grasslands of Lac du Bois, outside Kamloops from the chapter on California Bighorn Sheep, and the Swan Lake area of Princeton and Otter Lake region of Tulameen in the chapter on hummingbirds.

Other Area Interests

Great stops along the way to Manning include the loop trail at Bridal Falls Provincial Park, Coquihalla Canyon Provincial Park, the Othello Tunnels in Hope, and the Manning Park trails at Sumallo Grove and Rhododendron Flats (particularly if it's mid-June and the rhododendrons are flowering). If you happen to be at Manning Park in early to mid-May, consider a side trip to Princeton for hummingbirds. You can find out more about Princeton's hummingbirds in the chapter on that species.

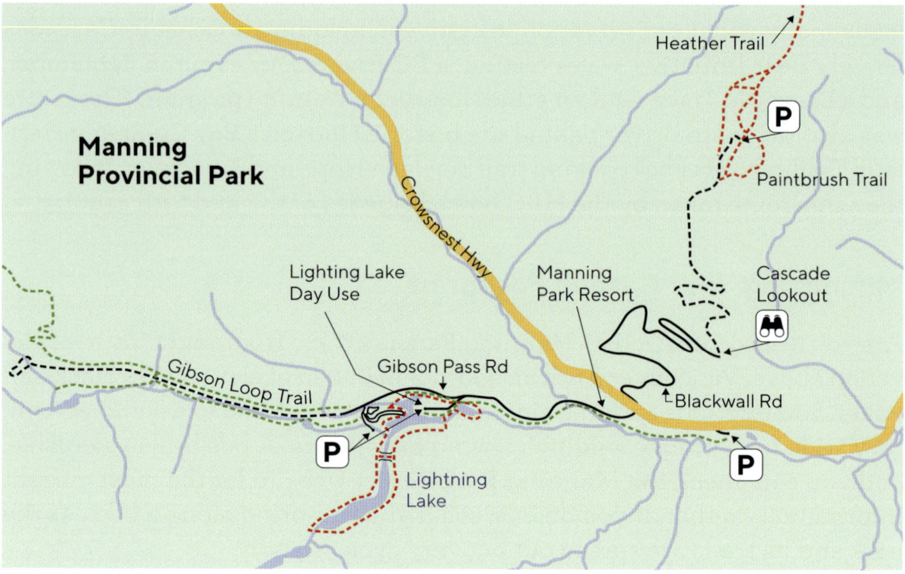

GOOD PLACES TO VIEW SMALL MAMMALS AT MANNING PROVINCIAL PARK INCLUDE LIGHTNING LAKES, CASCADE LOOKOUT, AND PAINTBRUSH AND HEATHER MEADOWS TRAILS. SOURCE: BC PARKS, MANNING PROVINCIAL PARK.

Additional Resources

1. **Small mammal track identification** is a good skill to learn, especially when you are in areas that are sandy, muddy, or in snow. I participated in a Nature Conservancy of Canada (NCC) webinar called "Summer Tracking for Beginners," led by Joseph Litke. It was fantastic and worth watching on YouTube as NCC keeps it available for public viewing.

2. **Hinterland Who's Who** is a historic Canadian nature educational program that is filled with great information on small mammals and much more.

3. Check out the ***BC Parks Wildlife Viewing*** brochure.

CLIMATE CHANGE AND SMALL MAMMALS

Changing climate can result in beneficial conditions in one season and difficult conditions in another, with potentially devastating consequences, especially for those eking out an existence in an already challenging alpine environment. Variable snow depth, mid-winter melts and refreezing, summer droughts limiting plant growth, and extended fire seasons all have ramifications for ground species that reside year-round in Manning Park, hibernating in place.

IT'S SPECIAL TO GET A CLOSE-UP OF A BAT FROM THE COLONY AT VASEUX LAKE, AN IMPORTANT AREA FOR BAT RESEARCH. JUST NEVER HANDLE THEM YOURSELF. LEARN MORE ABOUT BATS AT BCBATS.CA.
PHOTO JOHN SAREMBA.

Yuma & Little Brown Bats

in Peachland on Okanagan Lake

AT A GLANCE

WHAT: Bats – Yuma (*Myotis yumanensis*) and Little Brown (*Myotis lucifugus*)
WHERE: Peachland
WHEN: June–September (June, birthing; July, young begin to fly; mid-August, bats depart roost; September, mating; October, searching for winter hibernaculum)
TRADITIONAL TERRITORY: Syilx tmixʷ (Okanagan), Nłeʔkepmx Tmíxʷ (Nlaka'pamux), Okanagan
ACCESS: Easy

With all the myths and narratives we've created around them, bats must be among our most misunderstood and maligned species. Just as with wolves and sharks, our culture has created and exaggerated endless stories about them. These storylines begin to take shape when we are children watching Disney cartoons, strengthen through fiction in books, television, and film, and solidify when our adult imaginations take it from there. So, as you begin to read this chapter and consider a bat viewing experience, please set aside the following myths: bats get in your hair, bats are blind, all bats feed on blood, all bats carry rabies, bats aren't important or are a wicked enemy. False, false, false, false, and false! Let's begin a new narrative: Bats are interesting, important, threatened – and you can help bats. That's a comment from Ella Braden, a coordinator with the Okanagan Similkameen Conservation Alliance (OSCA) and a bat program volunteer. I spoke with Ella when researching this chapter, and these "true statements" can be accompanied by one more I discovered, which is that many people and communities have already stepped up to help bats. One of these is the Bat Education & Ecological Protection Society, BEEPS, in Peachland, BC.

JOIN THE BAT EDUCATION & ECOLOGICAL PROTECTION SOCIETY (BEEPS) IN PEACHLAND FOR AN EDUCATIONAL WALK ON THEIR INTERPRETIVE TRAIL. PHOTO DARLENE HARTFORD.

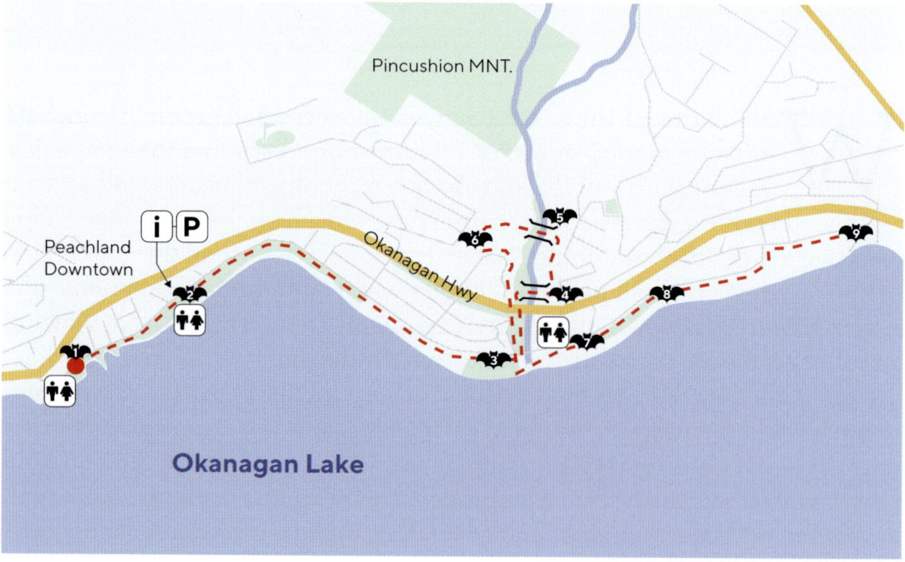

BEEPS HAS CREATED A BAT TRAIL WITH NINE INTERPRETIVE STOPS. IT'S A GREAT WAY TO SPEND AN EARLY EVENING IN PEACHLAND AND, WITH THE RIGHT TIMING, A GOOD CHANCE TO SEE BATS. SOURCE: BEEPSPEACHLAND.COM.

In 2019, Peachland became the second community in the province to be designated a bat-friendly community by the Community Bat Programs of BC. This means that Peachland has committed to protect, enhance, or create habitat, to provide bat information, and to promote learning about bats. Bats are loved by incredibly dedicated volunteers, local businesses, and the municipality, and that means Peachland is a fantastic location to learn more about these misunderstood species. The heart of Peachland's bat activity is at the visitor centre, a 1908 building with an attic that is home to about 2,000 bats from April to September (with a live video feed streamed into a public area where there are also educational displays). Community volunteers lead Bat House Interpretive Trail walks, give talks in local schools, do provincial bat counts every two weeks in summer, and promote bat-friendly LED lighting and gardens. Did you know that planting white flowering or silver foliage native plants in your garden attracts bat prey species such as moths and night flying insects? BEEPS provides a long list of plants that produce nectar late in the day, such as Lavender, Pearly Everlasting, Yarrow, Mock Orange, and Milkweed, which attract night pollinators. BEEPS also encourages home and business owners to consider green roofs with neonic-free plants. (**Neonics** are pesticides used to kill insects. Research shows they are also harmful to insect-eating wildlife.[68]) BEEPS's goal is to increase maternity, hibernation, and foraging habitat, as well as the availability of insects bats feed on. A walk on the interpretive trail will teach you about topics as diverse as:[69]

→ **Torpor** – Bats sleep for about 19 hours a day, foraging twice each night around dusk and just before dawn. In torpor, which is a deep sleep-like condition, bats' heart rate, metabolism, and body temperature all decrease. It's like a mini-hibernation that allows them to save energy. The attic-dwelling Peachland bats are believed to hibernate across Okanagan lake in the cliffs of Okanagan Mountain Provincial Park from October to March, living only on the fat they have managed to store over the summer months.

→ **Echolocation** – This radar system, using sound waves to bounce off objects and "read" the reflective echo, allows bats to navigate and find prey. The reflective sounds and their timing provide the bats with a sound "picture" of what is in front of them, its size, and how fast it's moving.

→ **Bat boxes** – These chambered structures can fit up to 300 bats, and some are outfitted with guano (bat poop) collectors, which volunteers and scientists use to monitor bat health and understand dietary preferences.

THE LONG-EARED MYOTIS EATS BEETLES ON THE GROUND AND ON TREES AS WELL AS AERIAL-HAWKS FOR INSECTS. BAT RESEARCH IS CRITICAL FOR UNDERSTANDING THE HEALTH OF BATS, WHICH ARE OFTEN UNFAIRLY MALIGNED IN THE EYES OF THE PUBLIC.
PHOTO ALAN BURGER.

BC is home to 15 of Canada's 19 species of bats. There are approximately 1,300 species of bats globally, which means *a fifth of the species of mammals on our planet are bats*![70] In BC only five species take advantage of bat houses, though, and therefore these provide only a partial answer to the needs of our endangered bats. The bigger solution is habitat creation and insect preservation. Both the Yuma and the Little Brown bats in Peachland use these bat boxes, which allows them to be counted by public volunteers each summer as they leave their boxes for night foraging (see Taking Action below).

→ **White-nose Syndrome** – This is caused by a fungus that thrives in cool, moist environments. It has been devastating to bats in many parts of North America, sometimes eliminating more than 90 per cent of populations, and it's the reason why the Little Brown Bat is endangered in Canada. But, thankfully, white-nose syndrome has not yet reached BC. Infecting the skin of hibernating bats, white-nose syndrome causes infected bats to wake up more frequently and therefore use more energy, often leading to death. Asking people to sanitize shoes and other outdoor equipment after they've been in caves with active bat populations is one

NATURALISTS JUDE AND AL GRASS HOLD A BAT BOX BEFORE INSTALLATION. BECOMING A NATURALIST IS A LIFELONG EDUCATIONAL ENDEAVOUR THAT BOTH BROADENS AND DEEPENS YOUR UNDERSTANDING ABOUT CONNECTIONS IN THE NATURAL WORLD. FORTUNATELY, NATURALISTS LIKE AL AND JUDE LOVE TO SHARE THEIR KNOWLEDGE. PHOTO ROB ALEXANDER.

important way to reduce transfer of the fungus from one site to another. Dr. Cori Lausen from the Wildlife Conservation Society Canada is testing a probiotic, bacteria-laden powder designed to fight the deadly fungus. Dr. Lausen and her colleagues hope to spread the powder around entrances to bat roosts to increase the density of existing good bacteria on bat wings, helping the mammals grow the microbes that will keep the fungus from developing.[71]

→ **Cool facts** – The Bat House Trail is bursting with cool interpretive facts that can create an opportunity to build engaging discussions with children. These include:[72]

- Bats feed on 600–1,000 insects per hour each night by "scooping" insects toward their mouth with their wings. Peachland's bat colony is a maternity roost, and reproductive females can consume up to 110 per cent of their body weight in insects *each night*!

- Predators include owls and snakes, but our cats eat bats too, when we let them out at night. If they can escape predators, bats may live for 20 years in the wild and up to 40 years in captivity.
- Wildlife trees are critical habitat for roosting bats, especially those near water. The Okanagan has lost many large Ponderosa Pines, potential wildlife trees, to vineyards, orchards, and waterfront homes.
- Bats provide an ecosystem service. By eating mosquitoes, moths, flies, and other insects, they are natural pest controllers, protecting our food crops (e.g., certain moths eat apples) and forests. Bats help cycle nutrients throughout the ecosystem (see regulating service below) by spreading their droppings, which are rich in nutrients like phosphorus and nitrogen and are important for tree growth. The guano is also becoming an increasingly popular fertilizer for local gardens. It is collected from the visitor centre's attic annually and, in the past, was even packaged and sold to the public.

The concept of **Ecosystem Services** is an idea best explored in this chapter in more detail, as it is very pertinent to both bats and the insects they feed on. If you scan the index in this book, you will see this concept appear in various places where it's clear that certain species, as well as the natural

BEEPS TAKES THEIR BAT EDUCATION PROGRAMS INTO LOCAL SCHOOLS, INSTILLING AN INTEREST AND A LOVE OF BATS INTO CHILDREN, AND IN TURN HELPING TO ENSURE THAT THE VARIOUS SPECIES ARE UNDERSTOOD AS VALUABLE MEMBERS OF THE OKANAGAN REGION. PHOTO DARLENE HARTFORD.

A BAT BOX AT COLONY FARM IN COQUITLAM IS FILLED "WALL TO WALL" WITH MATERNITY BATS HANGING TIGHTLY TOGETHER INSIDE FOR WARMTH AND PROTECTION. BOTH COLONY FARM AND DEAS ISLAND (IN DELTA) HAVE SIGNIFICANT BAT COLONIES. NOTICE THAT A FEW OF THE BATS HAVE SILVER BANDS ON THEM, PART OF THE RESEARCH AROUND WHITE-NOSE SYNDROME. *IF YOU HAVE A LARGE NATURAL AREA OR WATER BODY BY YOUR HOME, CONSIDER A SUNSET WALK TO LOOK FOR BATS.* PHOTO JOHN SAREMBA.

environment itself, provide benefits to people. In the section about pollination by hummingbirds, you'll see that pollination is considered a **Supporting Service**. Insects, bats (though not BC bats), bees, and hummingbirds are just a few of the species that help to pollinate plants, and in doing so offer a supporting service free of charge. The Ponderosa Pines that offer habitat for bats are providing a **Provisioning Service** for us in that they offer wood and fibre, also free of charge. The natural landscape, consisting of soil, shrubs, and plant roots, reduces water runoff and therefore acts as flood control. It is providing a **Regulating Service**, as is the cycling of nutrients dispersed through bat guano. Bats, as insectivores, also provide a regulating service as pest and disease control. Finally, bats and insects provide a **Cultural Service,** offering us educational and recreational opportunities as well as aesthetic pleasure that would be missed if they didn't exist. We don't pay to have these services provided, but if we had to, what would they cost?

THE HISTORIC PEACHLAND SCHOOL IS NOW USED AS A TOURISM VISITOR CENTRE. THE ATTIC IN THE BUILDING IS HOME TO A SIGNIFICANTLY LARGE MATERNITY BAT ROOST.
PHOTO DARLENE HARTFORD.

Pollination, for example, is done by hand in some places on Earth where pollinating species have been decimated, and it's a laborious, time-consuming, and very expensive undertaking. Considering the services nature provides and that we rely on, a deeper understanding of this critical connection is something we all need to take more notice of. So, if we value bats, then we really must value the ecological services bats in the Okanagan provide. An important counterpoint to the concept of ecosystem services is that it commodifies nature by putting a dollar value on nature. It's very anthropocentric. As a priority, we should be protecting nature for nature's worth, not just protecting it for its economic value to us.

Just as plankton is a foundational component of the ocean food webs, insects are a foundation of terrestrial **food webs**. BC bats feed only on insects, and they require massive amounts of them to thrive (see one of the cool facts above, regarding reproductive females), but our society spends massive amounts of money to eliminate insects from our agricultural environments in order to protect our crops. Cropland may appear to be wonderfully green with rolling hills from our perspective, but it can be a virtual desert, little different than concrete or pavement, to a bat who is unable to find any nourishment from it. Research from Germany shows a decline of insects in nature reserves across that country, on the order of 75 per cent in just 27 years.[73] The researchers noted that the speed and scale of the decline was shocking. Insects turn plants into protein and, when consumed, power the growth of bats and countless other species, such as freshwater fish and most birds, as well as all the creatures that eat these species. Just consider the Grizzly Bear that relies on the bees to pollinate the berries it eats, and the flies that feed the salmon that sustains it through hibernation. Many consider the worldwide insect decline apocalyptic, and you may observe this yourself on a drive to Peachland – just watch your windshield. Forty, or even 20 years ago, a drive to the Okanagan would ensure stops at service stations to clear the smeared insect carcasses off your windshield. Just ask an Elder. Today, however, there are far fewer bugs. This is known as the **windshield phenomenon** and helps us comprehend the insect deserts our BC bats are encountering. So, if insects and habitat are two of the crucial drivers that impact bats, you may ask if there are any others. And the answer is yes: one more big one, and that is artificial light.

Have you seen any of the NASA images of the Earth at night that show human-produced artificial light? It's astounding to see how far our night lights reach, and for night species like bats it really is a disruptive form of sensory pollution. In the presence of light, nocturnal animals alter behaviour patterns, breeding cycles, and even their hormone levels. A paper published in 2020 on the biological impacts of artificial light at night outlined how pervasive the effects are on organisms from microbes to invertebrates to animals and plants.[74] Furthermore, as we switch our homes and cities to LED lighting, a wider-spectrum luminosity more similar to sunlight, we risk even greater impacts. The authors write, "We need to start thinking about lighting in the way we think of other big systemic pressures like climate change." On the positive side, less illumination means less electricity, less cost, and fewer emissions, but it also requires a change of mindset and more thoughtfulness about our use of light. The authors of the paper continue, "We need to think in terms of using light only when we need it, where we need it and how we need it." Peachland, as a bat-friendly community, is moving in this direction, and the more they do, the more the bats will thank them for it!

We are learning a lot about where bats choose to roost in reference to humidity and temperature. It helps us determine where to put bat boxes. We look at environmental conditions, such as the importance of afternoon shade. When environmental conditions are poor, they are subject to pneumonia and heat stress. Currently there is no White Nose Syndrome in BC, but it will come and when it does, it will be devastating. So, it's extremely important to sanitize our shoes, outdoor equipment and avoid entering known bat roosts, when you are aware of an active bat population.

—**ELLA BRADEN,** Okanagan Regional BC Community bat program coordinator

Engaging and Taking Action

There are many ways to get involved with bats, including learning more through the BEEPS "Adopt a Bat" program, participating in a bat count, creating or enhancing habitat for bats, lobbying for a bat-friendly community, or volunteering your skills for projects such as bat house building. Bat counting in particular is a great family event to schedule into your Okanagan vacation.

ON MANY SUMMER EVENINGS BEEPS ASKS FOR VOLUNTEERS TO HELP COUNT LITTLE BROWN BATS LEAVING THE ROOST. ***CONSIDER THIS ACTIVITY AS AN OPPORTUNITY TO HELP YOUR FAMILY ENGAGE IN BETTER UNDERSTANDING THIS SPECIES.*** PHOTO DARLENE HARTFORD.

Every Friday in July and August, BEEPS holds a bat count. Locate their Facebook page for dates, times, and availability. In addition, Sunoka and Okanagan Lake Provincial Park are locations where you can volunteer as a citizen scientist to count bats. It's a great event for all ages and especially ideal for youth over 12. If you do a count, expect to be out at sunset, heavy with anticipation with a counting phone app or clicker in hand and counting bats as they literally drop out of a bat box. It's great fun but also important work, as counts help determine a baseline of data, something that is currently lacking with BC bat species. In the parks mentioned, counts are done before bat pups can fly, between the first and third week of June, and after they exit the roost between mid-July and early August. Counts begin at sunset and last about an hour. More information and instructions can be found at BCbats.ca. If you are on Okanagan Lake on a night with a deep blue sky that hasn't yet turned black, use a flashlight to shine over the water and watch for the Little Brown Bat skimming over the lake feeding on insects, silhouetted against the sky. This is truly a memorable summer observation.

If you choose to build or purchase a bat house, there are several considerations to take into account, including solar exposure, surrounding habitat, mounting height, obstacles, and protection from predators, lights, and wasps. **BCbats.ca** is the best place to learn more, and there is an opportunity on the website to communicate with a bat expert. Many BC naturalist clubs have ongoing citizen science projects to help support bats and their habitat. Find out what is being done at your local naturalist organization or consider going outside in your neighbourhood to monitor there. Even knowing there are no bats in an area is important information!

BEEPS has partnered with Hainle Winery on an agro-tourism project whereby BEEPS volunteers spread bat guano at the base of grape vines, as Hainle has found that the nutrients in the guano are a fantastic fertilizer for their grapes. A Bat and Vine walking tour is in the planning stages to educate people on the values of organic farming and the use of guano as a natural fertilizer.

Minimizing Your Carbon Footprint

Peachland is a small community with a lakefront trail connecting much of it. From Vancouver there are several bus companies that travel between Vancouver and Kelowna, and BC transit has service to Peachland from West Kelowna. Due to the proximity of services in Kelowna, this is one of the more easily accessible locations for leaving your vehicle at home and observing these bats and this community on foot.

ARE YOU A GARDENER? ***CONSIDER BAT-SPECIFIC SPECIES FOR YOUR GARDEN, SUCH AS WHITE FLOWERING OR SILVER FOLIAGE NATIVE PLANTS.*** **THESE ATTRACT DUSK-TO-DAWN INSECTS FOR THE BATS TO FEED ON.**

Directions and Accessibility

Highway 1 to Hope, Highway 5 to Merritt, and then Highway 97C to Peachland. The Bat House Interpretive Trail begins at Heritage Park on Beach Avenue. The visitor centre is at 5684 Beach Avenue.

The Bat House Interpretive Trail is a mixture of pavement and rough gravel and mostly accessible. Contact the Peachland Tourism Visitors Centre for more information.

Other Locations to View Bats

1. Consider joining an event where counts are happening, echolocation detectors are being used, or mist netting is taking place, such as during the Meadowlark Nature Festival (see the chapter on the Meadowlark Nature Festival). International Bat Week is another occasion with multiple events, and it takes place in late October.

2. Fintry, Johns Family Park, the cliffs of Okanagan Mountain Park, and Vaseux Lake in the Okanagan are all great places to see bats, as is Lightning Lake at Manning Park.

3. Delta's Burrvilla on Deas Island and Coquitlam's bat mini-condo at Colony Farm are both good viewing locations in Metro Vancouver.

Other Area Interests

Wanting to stretch your legs on a hike to a view? Check out the trails on Knox Mountain, where you can create a 7.5 km, 300 m elevation loop overlooking Kelowna and Okanagan Lake.

Additional Resources

1. COSEWIC Recovery Strategies are in place for many species including the endangered Little Brown Bat. Consider learning more about these at the COSEWIC website.

2. The Fraser Valley Regional Library loans out a **Bat Detection Kit** that helps you learn how to monitor and record bats. It comes with an echo meter that allows you to record bats flying above you. Check out FVRL.bc.ca/batpacks.

IT'S SPECIAL TO GET A CLOSE-UP OF A BAT FROM THE COLONY AT VASEUX LAKE, AN IMPORTANT AREA FOR BAT RESEARCH. JUST NEVER HANDLE THEM YOURSELF. *LEARN MORE ABOUT BATS AT BCBATS.CA.* PHOTO JOHN SAREMBA.

3. Take part in the Canadian Wildlife Federation's Bat House Survey that looks at bat box use and its internal climate to better understand the effect these structures have on bat physiology, survival, and reproduction.

4. Found a dead bat or have bats on your property? Or would like to build your own bat house? Go to **BCbats.ca.**

5. **Bats Out of the Darkness** is a travelling Kelowna Museum exhibit. Check in with the museum to see where it will be next.

6. If you are interested in developing and promoting bat habitat, see the ***USA Fish and Wildlife Bats*** brochure.

CLIMATE CHANGE AND BATS

Rapid and unpredictable changes in seasonal weather patterns, like those associated with climate change, can make it difficult for bat populations to survive. Heat waves and droughts can cause bats to die from heat stress or from lack of food. Massive wildfires, fuelled by the hot and dry conditions seen in BC in recent years, can destroy bat habitat. Intense storms and heavy rainfall can flood roosts, killing entire colonies of bats, while early or prolonged freezing temperatures can cause roosting bats to freeze during hibernation.

IT IS SUCH A POWERFUL EXPERIENCE TO MAKE EYE CONTACT WITH, AND BE SEEN BY, WILDLIFE.

PHOTO ROB ALEXANDER.

Urban Coyotes & Other Wildlife

in Metro Vancouver

AT A GLANCE

WHAT: Urban Coyotes (*Canis latrans*)
WHERE: Vancouver parks, fields, ravines, and golf courses. Especially Stanley Park or Pacific Spirit Park.
WHEN: All year
TRADITIONAL TERRITORY OF METRO VANCOUVER AND FRASER VALLEY: səl̓ilwətaʔɬ təməxʷ (Tsleil-Waututh), Skwxwú7mesh-ulh Temíx̱w (Squamish), sq̓əc̓iy̓aʔɬ təməxʷ (Katzie), Kwantlen, Coast Salish, sc̓əwaθenaʔɬ təməxʷ (Tsawwassen), S'ólh Téméxw (Stó:lō), Á,LEṈENEȻ ȽTE (W̱SÁNEĆ), Semiahmoo, Kwikwetlem, xʷməθkʷəy̓əm, Stz'uminus, Nuxwsa'7aq (Nooksack), Dënéndeh, Qayqayt, Ktunaxa ʔamakʔis, Hul'qumi'num Treaty Group
ACCESS: Varies

If you've ever had a Coyote make eye contact with you, or have heard the distant, haunting calls of a Coyote family yelping, barking, whining, or howling, you may have an inkling of what I'm referring to. A Coyote's stare and vocalizations seem to come from a place far away in time, which may explain why Coyote is known as a Trickster in the traditional stories of many BC First Nations, such as the Secwépemc and Syilx, as well as many of the Indigenous Peoples throughout North America.

Despite their prevalence in stories, and unlike other wildlife in this book, Coyotes may be more challenging to find in person. Urban Coyotes are often nocturnal (primarily active at night), or **crepuscular** (primarily active at dusk), although they can still be active in the daylight hours, but they are usually quite successful at hiding from humans.[75] Coyotes are

STANLEY PARK ECOLOGY SOCIETY

THE STANLEY PARK ECOLOGY SOCIETY (SPES) TAKES A LEADERSHIP ROLE IN EDUCATING THE PUBLIC ABOUT WAYS TO COEXIST WITH COYOTES. COURTESY OF THE STANLEY PARK ECOLOGY SOCIETY.

an "**intersectional**" species; that is, they inhabit a niche where urban and wilderness meet. They are animals that live in the "background" all around us, moving among the diversity of landscapes in our city, but if we don't actively seek them out, it's unlikely we'll catch more than a glimpse of them. That said, respecting this coexistence feels important.

I had the opportunity to interview Kristen Walker, an assistant professor in the Faculty of Land and Food Systems at the University of British Columbia (UBC), and referring to this coexistence she said, "It's like hey, I see you, and you see me, and we're doing our own thing, and that's okay." Kristen, a wildlife biologist, has a deep love for animals, and she sees each as an individual. Her desire is to treat wild animals more humanely and more compassionately, and that starts with how we humans respond to them.[76] It was important for me to ask Kristen, as a Coyote researcher in the Animal Welfare Program at UBC, if it would be responsible of me to include Coyotes in this book. My question was, should I be leading people to seek out Coyotes for the purpose of viewing them? Kristen's answer was that if we are to coexist with Coyotes in our neighbourhoods, then education is critical, as misinformation surrounding Coyotes abounds. So yes, we're going to discuss Coyotes, but pay attention to the fact that Coyotes are one of only a few species in this book where I suggest the animal as an experience for viewing, while simultaneously relating the latest best practices in research to deter, or aversion-condition, an animal that shares high-use areas with humans.

A good place to begin is with an organization that Kristen partners with: Stanley Park Ecology Society's (SPES's) "Co-Existing with Coyotes" program, which was founded in cooperation with the Vancouver Park Board and the BC Ministry of Environment. SPES set up the program to help *reduce conflict* between people, pets, and Coyotes. The program has run for two decades and is the longest-running Coyote coexistence program in North America. Their website allows the public to report local sightings, which means they're able to notify people in real-time about areas that Coyotes are currently frequenting, with the purpose of education that encourages a healthy coexistence between us and them. Areas in Metro Vancouver where Coyote sightings are common include most of the region's remaining habitat fragments such as Pacific Spirit Park, Stanley Park, and Queen Elizabeth Park. Other green fragments include ravines or wildlife corridors, golf courses, cemeteries, rail lines, and conservation areas. Therefore, if you'd like to increase your chances of seeing Coyotes, you should choose a location they're known to frequent, set up a walking route in the area, visit repeatedly early or late in the day, and spend time paying close attention to the natural surroundings. If you do see a Coyote, stop, observe, and genuinely appreciate the beauty and naturalness of the animal from a safe distance, but if it shows any sign of aggression or extreme tolerance to humans, know how to respond, considering the safety of yourself and others (see "humane hazing" below).[77] In any case, report your sighting to SPES, allowing them to map the location. Please keep the idea of coexistence in mind when seeking out Coyotes, and always observe the ethical viewing guidelines covered in the introductory chapter of this book.

Coyotes were originally a prairie and desert species; they prefer open areas (so our fragmented urban landscape works for them), but they also favour wooded patches and shrubbery that allows them shelter from people or other predators. They expanded their range into southern BC in the 1930s, typically choosing areas around farming communities. It wasn't until the 1980s that it became apparent they were inhabiting the cityscape of Metro Vancouver in larger numbers, so public perceptions of sharing our neighbourhoods with them are also relatively new. Through surveys and community dialogue over many years, SPES has discerned that most residents do not actually want Coyotes trapped and removed, but instead wish to find ways to coexist. But living alongside large, predatory animals in a city is complex and multifaceted, as there is a diversity of understandings, values, and preferences among residents as to how they relate to wild animals. Coyotes may look like dogs, but they're wild animals. So how do we coexist in a positive way? While some people remove attractants such as outdoor food dishes for pets or fallen fruit from trees, or lock composting bins, others do the opposite, putting out food to attract wildlife. Some people may hope to "commune" with the animals, but attracting a large, wild predatory animal is *always* a bad idea, which often leads to the death of the animal they are trying to reach out to through habituating behaviour to receiving food that is not their typical diet. Therefore, an effective coexistence framework that focuses on education, prevention, and ultimately investigation and enforcement of safety bylaws is required.

My attempt to understand this more fully led me to contact the Urban Wildlife Programs Coordinator at SPES, Dannie Piezas. In the summer of 2021, Dannie was overwhelmed with reports of aggressive Coyotes on the

NOTICE THE TEETS ON THIS COYOTE MOTHER, WHICH IS LIKELY STILL NURSING. FOR VIEWING COYOTES WITH THE LEAST IMPACT, *BRING BINOCULARS TO ALLOW THE GREATEST DISTANCE POSSIBLE BETWEEN YOU AND THE ANIMAL*. PHOTOGRAPHS LIKE THIS ARE TAKEN WITH A ZOOM LENS. PHOTO ROB ALEXANDER.

Stanley Park Seawall. The issue got so bad that the BC government had to step in to do a Stanley Park–wide Coyote cull. Part of the problem was that people were putting food out for the Coyotes. That is the antithesis of coexistence with urban wildlife. Therefore, Dannie wanted to share with me the concept of "**humane hazing**" as it relates to Coyotes. If a *Coyote is approaching* you, your property, or your pets, the proper response is hazing. Also known as *aversion conditioning*, the idea is to provide people with a reliable and non-lethal "set of tools" to generate a *negative association* between the Coyote and themselves, ideally creating a healthy boundary between humans and Coyotes.[78] The way to do this is to communicate to the Coyote that it *must leave*, so look big, wave your arms or a stick, shout or make noise (some people carry homemade noisemakers, like a pop can filled with rocks), or open and close an umbrella; but don't run away, as that could instigate their chase instinct! Sealing off open spaces under buildings or decks, and securing garbage or compost containers, are relatively simple adjustments you can make if a Coyote is frequenting your property. And if it's becoming too comfortable, hoses or sprinklers can be a significant deterrent as well. Like other large mammals, Coyotes stand their ground or are confrontational with dogs, especially when safeguarding their family or their territory. Since we have the ability of foresight to manage our actions around our pets, we should try to keep conflicts from happening, not just for Coyotes but for other wildlife as well. Controlling our own actions and behaviours, as well as learning to understand Coyote behaviour, goes a long way toward ensuring our successful coexistence.

January to February is Coyote courtship and mating season, and their wider breeding season includes claiming a territory, denning, and, shortly thereafter, birthing a litter of generally between four and seven pups. When denning areas are known, it's important to temporarily restrict dogs by leashing them until the end of April, in order to have more control of your pet and to protect the Coyote pups. Dens are usually made in hollowed-out tree stumps, in crevices of large rock outcrops, or in existing burrows created by other animals. A Coyote pack typically consists of five or six adults, and rearing of pups *by the pack* continues through to the end of August. Interestingly, only the alpha pair mate, while the subordinates usually help to raise the young. Coyotes are particularly family oriented, and pack structures are essentially the parents and young (both older siblings from previous years and pups from that year). Many juveniles begin hunting on their own through the fall and early winter. Within one year, the pups are

THOUGH THEY ARE RARE TO SEE, WATCHING THE POUNCING HUNTING STRATEGIES OF A COYOTE MAY HELP YOU UNDERSTAND THEIR INCREDIBLE SENSES, WHICH HELP THEM TO BE ACUTELY AWARE OF ALL ASPECTS OF THEIR ENVIRONMENT, EVEN THOSE OF RODENTS MOVING UNDERGROUND. PHOTO MILVA DESIENA. SOURCE: STANLEY PARK ECOLOGY SOCIETY.

fully grown, and many disperse to find new territory. Therefore, in urban areas it's not uncommon to encounter young, solitary Coyotes that have left packs and are looking to join groups or to claim their own territory.

Metro Vancouver's population of Coyotes is stable and, overall, healthy. Two additional and significant challenges to a healthy population, however, are vehicle collisions and the use of **rodenticides**. The most common cause of death for urban Coyotes is collisions with vehicles (40–70 per cent of annual deaths) as they regularly cross roads while traversing their territories.[79] As for homeowners using rodenticides, I'd suggest that before considering using this, check to see if your city or municipality is reviewing the use of them (or if they are already banned). Although rodenticides are meant for killing rodent pests, the chemicals can enter the food chain, affecting other predatory species such as hawks, owls, and eagles as well. Most rodenticides are anticoagulant-based, which means they work as blood thinners, causing animals that ingest them to die from excessive bleeding. Rodents that have ingested this poison may be weak, but not dead, and therefore easy prey for an unsuspecting Coyote; and since rodents make up about 80 per cent of their diet, the Coyotes are likely to ingest and suffer from the rodenticide effects as well. As a major rodent predator, Coyotes provide an *ecosystem service* of natural pest control, but with rodenticides, providing that service comes at a cost to them.

Coyotes are prime examples of **ecological resilience** and **adaptation** in our cities, but our behaviour around them still affects their overall health and habits. They are an extremely intelligent omnivore and can learn quickly, conforming their behaviour to the circumstances, while diversifying their diets. Besides small mammals or birds with eggs (e.g., geese), they can also be found eating berries, fruit, nuts, fallen seeds from our bird feeders (so please keep these clean), compost, and garbage. But urban Coyotes' predilection for small urban (and un-streetwise) dogs and cats places them on the anger-radar for pet-loving residents who feel the conflict magnified when their own pet disappears in the night. The Urban Coyote Research Project, based out of Ohio State University, is a leading authority on Coyote research, and their data collection indicates that pets only account for 1–2 per cent of a Coyote's diet.[80] But with population estimates of 2,000–3,000 Coyotes in Metro Vancouver, that predation can lead to many lost pets! It might surprise you to know that pet owners on Vancouver Island or the Gulf Islands don't have this concern, as Coyotes have never made it to the islands!

If humans were to *honour* the sharing of space, how might that be represented? I'd like to suggest it could look like bioregional mapping. I first learned about bioregional mapping from Doug Aberly, a "bioregionalist" mapmaker and professor in UBC's School of Community and Regional

AS OMNIVORES, COYOTES SOMETIMES HUNT SMALL ANIMALS NEAR OUR HOMES, WHICH ARE PREDOMINANTLY RODENTS BUT CAN INCLUDE OUR UNSUPERVISED PETS.
PHOTO MICHAEL SCHMIDT PHOTOGRAPHY. SOURCE: STANLEY PARK ECOLOGY SOCIETY.

Planning. **Bioregional Maps** are made by citizens to describe a community's understanding or perception of itself – its identity – through a thorough recording of pertinent biophysical and cultural information. Elements that are mapped might include wildlife and their corridors and habitats, wetlands, environmental hotspots, watersheds, geology, how First Nations used (and use) particular areas, and First Nations, claims and distribution. Map content is collected by visiting real-world places, talking to Elders in the community, taking photographs, interviewing community members, and recording experiences.[81] Ultimately, the goal is to develop a sense of place and connection to the land in a more holistic way. For example, when I consider Coyotes in my neighbourhood, I note the places I see them, I consider the habitat where their dens may exist, and I pay attention to potential overlaps with people that may lead to conflict. This becomes a layer in the mental bioregional map I carry in my head of my own urban landscape in the Metro Vancouver community where I live. You can imagine what your map would look like as you consider salmon in your local creek, nesting herons, or a First Nation's "**midden**" that shows thousands of years of occupation.

COYOTES MAY CREATE THEIR DENS AND RAISE THEIR FAMILIES NEAR URBAN AREAS, SUCH AS THIS ONE WITH SEVEN PUPS AT VAN DUSEN GARDENS. PHOTO ROBYN WORCESTER. SOURCE: STANLEY PARK ECOLOGY SOCIETY.

Coyotes are just a fraction of the urban wildlife that call Metro Vancouver home. Skunks, Racoons, rats, squirrels, and other animals inhabit our urban and suburban landscapes, and each has characteristics that require different wildlife control-measures (*and people control-measures!*) that, when implemented widely, can also lead to friendly coexistence. The BCSPCA and WildSafeBC provide information-rich resources for living with urban wildlife, and the latter also has a Wildlife Alert Reporting Program with a map for large mammals, such as Coyotes, that are seen in our neighbourhoods. A key reason *why* we all should learn about measures we can take to coexist happily with animals is so we can *reduce human-wildlife conflicts*. If you search the BC government environmental protection and sustainability website, you'll come across Predator Conflicts & Statistics. Just a cursory scan of the data clearly shows the annual number of conflict encounters (for bears and Cougars) that BC conservation officers attended, along with the numbers of animals destroyed, translocated, or hazed. The numbers destroyed are significant and show there is still much room for education and behaviour change on the part of humans.

Most Metro Vancouver communities have websites that provide information on urban wildlife, including: Coquitlam (Wildlife Management), Surrey (Wildlife & Habitats), Vancouver (Urban Wildlife Control), North Vancouver (Living Safely with Bears and Other Wildlife and Wildlife Awareness), Maple Ridge (Wildlife Safety), Richmond (Urban Wildlife), and Port Moody (Wildlife). Most communities suggest that if you come across an injured animal, you should contact the Wildlife Rescue Association of BC. But back to that "*why*." Dan Straker, a former SPES urban wildlife programs coordinator, provided me with the crux of this when he said, "Coyotes forced me to unlearn what I thought about as nature and natural, and about the separations we make between ourselves and nature. Coyotes are part of wilderness in our city. Nature and wilderness are in us, and we are part of them. This was an epiphany for me."[82] Finally, Dan reminded me of a quote from poet and environmental activist Gary Snyder: "Nature is not a place to visit. It is home."

> Our actions and relationships are part of a bigger picture, and there's a deep fulfilment in both the learning of, and knowing how, I fit in. It helps give me personal meaning.
>
> —**DANNIE PIEZAS,** Co-Existing with Coyotes program coordinator

Engaging and Taking Action

Among the most important things you can do to help support Coyotes is to spread the ideas about coexisting with Coyotes and the existing programs dedicated to this. There are valid reasons people dislike Coyotes, especially if they have lost a pet and believe Coyotes were responsible. Negative opinions and controversy can also lead to the spread of misinformation and fear peddling. Therefore, paying attention to the possibility that misinformation can exist is important as well.

Trail and security cameras around people's homes are becoming more common. Though Kristen Walker plans to eventually use trail cameras to individually identify Coyotes through the use of artificial intelligence, you can set up your own camera to discover which animals utilize your property, and then share that knowledge with your neighbours as you segue into the realization that you are already coexisting with animals near your home. Trail cameras are a way of capturing an image of an animal without invading their space. Remember, Coyotes are not a species you should chase after, for photography or any other reason.

As with other species in this book, the more you can become actively involved in their protection, the better. Connect with your local naturalist club, Streamkeepers group, or the environmental programming running through the MyImpactPage at Metro Vancouver Parks, or your local community centre. Finally, inform yourself about, and potentially speak out on, rodenticides in your community, as they can lead to the poisoning of animals higher up the food chains, not just rodent pests.

OFTEN THE FIRST THING WE NOTICE ABOUT A COYOTE AFTER WE RECOGNIZE IT IS ITS APPARENT HEALTH, WHICH CAN RANGE FROM SCRAWNY AND SCRAGGLY TO FIT AND PLUSH, AS IS THIS COYOTE. PHOTO CHRISTINE GAIO.

Minimizing Your Carbon Footprint

One thing about urban wildlife species is that they reside in areas with abundant transit options and cycle infrastructure. In fact, I'd suggest that your opportunities to see urban wildlife increase when you follow cycle paths or urban walking trails away from traffic.

Directions and Accessibility

Both directions and accessibility vary depending on the green space you choose to visit. Consult the SPES Coyote Sightings or the WildSafeBC Wildlife Alert Reporting resources for current advice.

Other Locations to View Coyotes

Throughout Metro Vancouver, the Fraser Valley, and Okanagan in green space habitats that are surrounded by urban areas.

Other Area Interests

1. Pacific Spirit Park, an important Coyote habitat, happens to be near the Beaty Biodiversity Museum, which is Vancouver's main museum of natural history. It houses collections and exhibits, and runs many events, highlighting the flora and fauna of BC.

2. The Stanley Park Nature House is located on Lost Lagoon in Stanley Park and is operated by the Stanley Park Ecology Society. It has interpretive displays and is an important area to visit if you are wanting to delve into the natural history of the area, including the Coyotes of Stanley Park.

3. Other urban wildlife species in Metro Vancouver include Racoons at Lost Lake, and Beavers at Beaver Lake in Stanley Park. Beavers are also commonly seen at Burnaby Lake and in Minnekhada Marsh. Both Racoons and Beavers are more likely to be seen at sunset.

4. A little beyond Metro Vancouver, consider a walk to Crippen Regional Park on Bowen Island to see Black-tailed Deer. There is more information regarding a trip to Bowen Island in a previous book I wrote titled *Active Vancouver: A Year-Round Guide to Outdoor Recreation in the City's Natural Environments* (RMB/Rocky Mountain Books, 2015).

5. Though you're unlikely to see a Cougar in Metro Vancouver, they do inhabit the region. If you do encounter one and it expresses interest in you or your pet, don't run or turn your back. Maintain eye contact with it, look big, yell, be assertive, and back away slowly. Carry bear spray when in areas they've been known to frequent. Get more information at WildsafeBC.

Additional Resources

1. **Coyote Watch Canada** is a federal, non-profit, community-based, wildlife organization advocating for positive human wildlife experiences.

2. **BC SPCA** has a site that suggests many ways to help urban wildlife.

3. **Wildlife Rescue Association of BC** is the main location in Metro Vancouver to contact if you find injured wildlife.

4. **Ministry of Environment and Climate Change** is a government site with lots of good information on human-wildlife conflict.

5. Climate change will affect urban wildlife. **Climate 2050** is a climate strategy for the Metro Vancouver region. It is worth reading the strategic framework of this plan to see how the municipalities of Metro Vancouver are approaching climate change.

6. Learn more in a podcast from **CBC's Quirks and Quarks** titled, "How Coyotes have managed to find success in the city like no other predator."

BLACK-TAILED DEER ARE A COMMON URBAN SPECIES. A FAVOURITE PLACE TO SEE THEM IS FROM THE WALKING TRAILS AT CRIPPEN REGIONAL PARK ON BOWEN ISLAND. THEY ARE ALSO KNOWN AS MULE DEER BECAUSE OF THEIR LARGE EARS.
PHOTO SABINE DECAMP.

WHATEVER YOU THINK OF THE WILES OF RACOONS, YOU CAN'T HELP BUT BE TAKEN IN BY THEIR LOOKS AND THEIR CLEVER BEHAVIOURS. AS WITH OTHER URBAN WILDLIFE, *THE BEST WAY TO ENCOUNTER THEM IS BY FREQUENTING NATURAL AREAS, AND NOT BY ATTRACTING THEM TO OUR HOMES.*
PHOTO MICHAEL SCHMIDT PHOTOGRAPHY.

CLIMATE CHANGE AND COYOTES

Coyotes are a generalist species, eating a variety of foods and thriving in a range of habitats. This adaptability bodes well for their survival with a changing climate, and the fact that they find our cities so habitable may mean their populations will likely increase, bringing more potential for human-animal conflict. That is all the more reason for planners to make public education that encourages *successful coexistence* a priority.

FOR CRABS AND KELP TO SMALL FISHES, BRING A GUIDEBOOK TO HELP YOU LEARN AND SHARE WITH OTHERS ABOUT THE SPECIES IN TIDE POOLS. *BE CAREFUL TO NOT INTRODUCE SUNSCREEN TO THE POOLS.*

Intertidal Life & Tide Pools

at Botanical Beach, Juan de Fuca Provincial Park

AT A GLANCE

WHAT: Tide pools filled with intertidal life
WHERE: Port Renfrew
WHEN: Summer, especially at the low tides of the full and new moons
TRADITIONAL TERRITORY: Pacheedaht
ACCESS: Easy

Shifting our perspective can provide us with a valuable tool for exploring and appreciating the natural world. As Dr. Richard Kool expresses in his "Introduction to Summer" in this book, to lie on your stomach on warm sandstone, and gaze into a tide pool at Botanical Beach in Juan de Fuca Provincial Park, makes the ordinary world simply disappear. But what opens up is another world, filled with movement, competition, predation, and a myriad of species skulking about or lying in wait! The excellence of this site for viewing intertidal life is due to the presence of tide pools. The action of waves, wind, and pebbles has slowly worn away the sandstone over time, carving deep holes that fill up with sea water as the tide comes in, but don't empty when it lowers. The pools offer refuge to creatures – a refuge that protects its inhabitants from the living and non-living threats from both land and sea.

Consider the adaptations required for species of this environment to survive in this rugged habitat called the intertidal zone. Species of the intertidal zone experience extremes by several different measures. When the tide is out, they must contend with the heating effects of the sun, the drying influence of the wind, and with rainwater deluges, which can drastically dilute the tide pool seawater, to the detriment of its saltwater-adapted inhabitants. In addition, they must cope with predation by birds, mammals, and reptiles, all moving in to feed on the freshly exposed prey, revealed as

WHEN THE TIDE GOES, THE TABLE IS SET...EVEN FOR GULLS! IT TAKES PATIENCE TO RELAX A SEA STAR ENOUGH TO FIT DOWN A THROAT, BUT IT IS NOT UNCOMMON TO SEE.
PHOTO CHRISTINE GAIO.

the ocean's edge recedes. But when the tide is in, those same species must contend with the chilling effects of seawater, strong wave action, and ocean predators, which can now swim over and pluck them from their precarious homes. At all times, they're at risk from the other inhabitants of their tide pools, including sea stars, carnivorous snails, and crabs. When you look in the tide pools, spend some time pondering how the various species you observe may have adapted over time, improving their chances of survival in this variable ecosystem.

I had the pleasure of speaking with Phil Lavoie, a long-time educator at the Bamfield Marine Sciences Centre, about the health of coastal intertidal zones and how best to experience them. He says the best way to see an entire ecosystem of species is to "get your face right down there." Phil adds that if you look long and hard enough, you'll see that most pools have 20 to 30 different species in them![83] But before we take our "walk" and consider what organisms might exist in these pools, and how they're able to survive in this severe environment, it's worth considering some points on tide pool safety and etiquette:

→ First, consider when to visit and with whom, as this can definitely be a family viewing experience for all ages. Expect to walk 1.5 km on a forest trail from the parking area to Botanical Beach, and more around the intertidal zone. Your total time exploring will be dictated by the tides, so to maximize your time, it's critical to arrive at a falling or low tide, and the lowest of these in the summer happen around the full and new moons. Check out the Port Renfrew tidal prediction chart at tides.gc.ca.

→ Second, consider *safety*. Wear sturdy shoes or, better yet, rubber boots with good grip. Be aware of slippery seaweed, and when you're near the ocean's edge, don't turn your back to it as *random large waves* (known as "rogue waves" or "sneaker waves") do happen occasionally, and can sweep even the most sure-footed off their feet. If there are caves in the cliff faces, never approach them if the tide is rising as it's easy to become trapped. Remember, this is a wilderness area and first aid is not readily available.

→ Third, approach with the etiquette of a considerate wildlife viewer, meaning refrain from tearing off creatures that are attached to the rocks, and if you lift a rock to look at what might be hiding underneath, always replace it in the exact same position you found it. (Think of rocks like the roof of a home: Wouldn't you want your roof put back if someone removed it to look inside?) Do not remove, collect, or disturb any tide pool life, shells, plants, kelp, etc.; make sure your hands are *clean and free of lotions or sunscreen*; and even then only touch a tidal pool creature softly and preferably with just your baby finger. Also consider that your footsteps have the potential to squish delicate organisms, which can be a big threat to life in an intertidal zone, so tread as lightly as possible.

→ Fourth, don't forget to watch the ocean for Killer Whales (summer), Grey Whales (March–April), sea lions (August–March), seals (all year), or porpoises (summer), as they all live in the waters just offshore.

→ Finally, bring a tide pool guidebook with you, as well as a camera, and, if you happen to have a pair of binoculars that will focus up close or a jeweller's eye loupe, bring those too!

THE BAMFIELD MARINE SCIENCES CENTRE IS AN IMPORTANT COASTAL RESEARCH AND EDUCATION CENTRE. *IN PREPARATION FOR VISITING THE WEST COAST, BE SURE TO CHECK OUT EDUCATIONAL PROGRAMMING THAT YOU CAN PARTICIPATE IN*. PHOTO PHIL LAVOIE.

Oh, and now about that "walk" – let's go! After you exit the temperate forest trail, the intertidal area is laid out as a nearly flat table in front of you, and your first choice will be to go right or left: the right fork leads to Botany Bay, while going left takes you to Botanical Beach and more accessible tide pools. (Note that from the parking lot you walk to the beach and back again, *OR* you can do a 2.9 km loop between the parking lot, Botanical Beach, and Botany Bay, choosing your direction based on the timing of the tides.)

Seaweeds

Before you even get to a tide pool, you may recognize an array of seaweeds. Start by considering how these species adapt and survive here. The green sea moss has glass-like silica in its cell walls, which discourages other organisms from grazing on it. Most species of seaweed have a holdfast that anchors the plant to the floor of the ocean. Some have air bladders for

HEALTHY UNDERWATER BULL KELP FORESTS PROVIDE HABITAT AND PROTECTION FOR MANY SPECIES. *WHEN YOU FIND BULL KELP ON LAND, SEE IF YOU CAN IDENTIFY ITS HOLDFAST ON ONE END, ITS GAS-FILLED BULB ON THE OTHER, AND ITS FLOATING BLADES, WHICH ALLOW IT TO PHOTOSYNTHESIZE AT THE OCEAN SURFACE.*

floating, helping them capture the rays of the sun. Seaweeds are algae that capture carbon, photosynthesize, and produce oxygen, just like terrestrial plants. Many seaweeds are critical for supporting other species, both as phytoplankton, a seaweed's tiny, juvenile form (which is consumed by the tiniest of fish to the largest of whales), and as habitat when they are fully grown, just as an old-growth forest provides habitat for many other species on land. Seven hundred unique seaweeds are found between California and Alaska, and all are classified as either brown, red, or green.

A short checklist to look for and identify includes:

Sea Lettuce ✓
Rockweed ✓
Bull Kelp ✓
Sea Moss ✓
Sea Palm ✓

Sea Anemones

At the tide pool's edge, even before crouching down, you'll see floral-looking creatures called sea anemones, which are not flowers at all but rather animals. Their waving tentacles are covered in stinging cells to protect themselves, and they also function to catch and stun prey before funnelling it toward the mouth, which lies in the centre. Gently stroking those tentacles with your baby finger (free of lotion) is harmless, and the animal will quickly retract and disappear. Giant Green Anemones also harbour photosynthesizing green algae in a symbiotic relationship, giving them their green colour. Many anemones (especially Aggregating Anemones) are known to multiply asexually, dividing into identical individuals. This results in entire rocks being carpeted in anemone clones! As the tide goes out, the anemones living on the rocks will retract their tentacles, appearing to close up, and they sometimes end up covered with light-coloured shells, which could help to reflect some of the sun's heat and retain moisture.

A short checklist to look for and identify includes:

Giant Green Anemone ✓
Aggregating Anemone ✓

OFTEN MISTAKEN FOR FLOWERS, GREEN SEA ANEMONES ARE ANIMALS WITH STINGING TENTACLES, WHICH ALLOW THE CREATURE TO STUN SOFT-SKINNED PREY AND FUNNEL IT TOWARD ITS MOUTH IN THE CENTRE. *ALMOST EVERY TIDE POOL EXCURSION INCLUDES A SIGHTING OF ANEMONES OF SOME KIND!* PHOTO STAN SHEBS, WIKIMEDIA COMMONS, CC BY-SA 3.0.

PRIOR TO SEA STAR WASTING DISEASE AFFECTING THE BC COAST BEGINNING IN 2013, SUNFLOWER STARS WERE A COMMON SIGHT. TODAY, HOWEVER, THESE ARE RARE TO SEE AS MOST HAVE SUCCUMBED TO THIS DISEASE, LIKELY ONE RESULT OF WARMING OCEAN WATERS.

Sea Stars

Commonly called "starfish," these animals have a skeletal makeup of calcium carbonate plates, called ossicles. Known as invertebrates, they lack a spine and are called **echinoderms** (where "echino" = spiny, and "derm" = skin). They often have a tough upper layer and a soft underside filled with hundreds of tube feet, all controlled by *hydraulics*. These feet are used to pull open the shells of clams or mussels, and then the sea star extends its stomach inside the shelled creature to begin digesting it. Ochre Stars are considered a *keystone predator* in the intertidal due to the influence they have on other creatures living there, specifically controlling the spread of mussels. Unfortunately, Ochre Stars have been hit hard by a virus commonly known as **sea star wasting disease**, which has caused an 89 per cent decline in their numbers since 2013! There is still much to be learned about this devastating occurrence, but many think the warming of our coastal waters due to climate change has reduced the resilience of sea stars in warding off this virus. By summer 2021 local divers were relating their concerns of a similar wasting with Sea Cucumbers, another echinoderm. Stay tuned, and if you want to help, see Engaging and Taking Action below.

A short checklist to look for and identify includes:

Ochre Star ✓
Leather Star ✓
Blood Star ✓
Bat Star ✓

OCHRE STARS ARE ALSO CALLED PURPLE SEA STARS; THEY CAN OFTEN BE FOUND IN EITHER COLOUR. CONSIDERED A KEYSTONE SPECIES, SEA STARS EAT AND THEREFORE KEEP MUSSELS IN CHECK IN THE INTERTIDAL ZONE. PHOTO STEVEN PAVLOV, WIKIMEDIA COMMONS, CC BY-SA 4.0.

Sea Urchins

Like sea stars, sea urchins are echinoderms. These spine-covered animals that graze on seaweed are a primary prey of Sea Otters, and there is a strong relationship between seaweed, urchins, and otters such that otter numbers (or lack thereof) can have significant impacts on urchins and, in turn, on seaweed. The hard moveable spines are made of calcium carbonate and offer some protection, while on their undersides the urchins have tube feet, like sea stars, which they use for locomotion, sensing their surroundings, and moving food toward their mouth. Purple and Green Sea Urchins can live to 20-plus years, while Red Sea Urchins are known to be among the longest-living animals on Earth, living as long as 100–200 years![84] Watch for these deeper in the tide pools, often squeezed into crevices or in a self-excavated hole etched by their teeth into the sandstone.[85] Their roe (eggs) are a delicacy in many sushi restaurants. BC is home to three different species, each defined by their colour.

A short checklist to look for and identify includes:

Purple Sea Urchin ✓
Green Sea Urchin ✓
Red Sea Urchin ✓

PURPLE SEA URCHINS ARE ONE OF THREE COMMON SPECIES OF SEA URCHINS ON THE COAST OF BC, AND THE ONE MOST LIKELY TO BE SEEN DURING INTERTIDAL EXPLORATIONS. LOOK FOR THEM IN POOLS OR ROCK CREVICES DURING THE LOWEST TIDES. PHOTO ED BIERMAN, FLICKR, CC BY 2.0.

THERE ARE MANY SPECIES OF CHITON, AND THEY ARE FOUND ALL AROUND THE WORLD, BUT THEY HAVE ONE THING IN COMMON: THEY ALL HAVE EIGHT OVERLAPPING PLATES. THIS IS ONE OF THE MORE STUNNING ONES, THE LINED CHITON. PHOTO MATT KNOTH, FLICKR,

Shelled Creatures

Many species in the tide pools and the surrounding area form shells. Shells are built from calcium carbonate, but because the oceans are taking up approximately 30 per cent of human-generated CO_2 emissions, our excess carbon dioxide is steadily acidifying the oceans, thereby making the formation of shells more difficult.[86] Look closely at a tide pool to discover some of these shelled creatures such as limpets (a hat-shaped shelled mollusc that grazes algae on the rocks, and suctions tightly if you try to move it); chitons (another grazer with eight overlapping shell plates protecting its soft body underneath); snails (both herbivorous grazers of rock algae, and meat-eating snails that drill holes into other shellfish to consume the contents); and mussels and barnacles, the filter feeders of the intertidal, each with a different system for holding tight to the rocks (the former using hair-like byssal threads, and the latter by secreting an adhesive).

A short checklist to look for and identify includes:

Plate Limpet ✔
Lined Chiton ✔
Dog Whelk ✔
Periwinkle ✔
Acorn Barnacle ✔

Other Tide Pool Species to Identify

By now you have your face just above the tide pool and you're noticing movement in your peripheral vision: the slow-moving shell is likely the home of a Hermit Crab (an animal that doesn't make its own shell but acquires them from others), while the fast-darting fish is likely a Tidepool Sculpin, moving quickly between kelp strands, which provide places to hide. Also watch for Rockweed Isopods, four centimetre, greenish to brown scavengers of the intertidal, often clinging to kelp or rocks. Throughout your wider exploration, you're also likely to come across the giants of the intertidal: California Sea Cucumbers (soft-bodied echinoderms that are often found in rock crevices); Sea Squirts (also known as "tunicates"), often appearing as shiny colourful hanging blobs; or, my favourite, Gooseneck Barnacles, which you'll find out at the ocean edge where the strong waves crash against the shoreline, bending these long, rubbery, filter-feeding creatures – but they always seem to bounce back to meet the next wave!

If in your "tide pooling" experience you were enthralled by the interactions of Hermit Crabs, the identification of Isopods, or the slimy texture of a Sea Squirt, then I'd say you're a great candidate for further education. The Bamfield Marine Sciences Centre is a world-class education and research facility that provides "access to a remarkable diversity of marine, terrestrial, freshwater, and cultural sites of the North East Pacific basin."[87] Based out of Bamfield, the centre accommodates visiting researchers and offers university-credited courses, as well as non-credit field trip programs. Other opportunities for further marine education exist through West Coast N.E.S.T. (Nature, Education, Sustainability, Transformation), a hub convened through the Clayoquot Biosphere Trust that shares upcoming educational programs available in science, nature, history, language, and Indigenous culture. And if your journey takes you to Ucluelet, be sure to visit the volunteers and tanks of Canada's first collect-and-release aquarium. The Ucluelet Aquarium offers hands-on learning about the marine life of

BINOCULARS CAN BE VERY USEFUL FOR VIEWING INTERTIDAL LIFE, AS WELL. ***CHOOSE BINOCULARS THAT CAN FOCUS UP CLOSE, AS WELL AS AT A DISTANCE.***

Barkley Sound, and hosts an array of projects ranging from species and marine debris surveys, to beach talks, walks, and cleanups.

Do you know a youth between 18 and 30 who shows an interest in conservation of the ocean environment? If so, the non-profit Ocean Wise runs an 11-month coastal conservation program called Ocean Bridge for Canadian youth, empowering them to take action toward ocean protection. As an ambassador program, it involves community and capacity building, mentorship, and local service projects, as well as an annual "summit" to bring together past and present participants. They recruit 160 youth from across Canada and provide funding support for participants in their multi-day immersive learning journeys, in the form of transportation, meals, and accommodation, as well as adaptive technology, online access, and child care if required. As a university professor, I often hear from my students that they want to have a career that is purposeful and makes a difference in this world. Yet they often seem paralyzed in the present moment, caught up in the busy-ness of post-secondary coursework, and not able to easily see how to connect their interests to the bigger picture of global issues. Ocean Bridge helps make that connection, and it's one of many organizations that support youth in making a commitment to conservation.

Conservation of our ocean environments is critical for the long-term stability of the planet, which is tied to our own resilience. Oceans drive the global systems that make Earth habitable for humankind. They cover more than 70 per cent of the planet and contain 50–80 per cent of all living species. Oceans are the primary source of the oxygen we breathe, and most of this is produced by phytoplankton. Yet ocean warming, acidification, invasive species, overfishing, acoustic noise, plastic waste, and nutrient pollution that leads to dead zones in the ocean are all real concerns. I'm encouraged that Canada is one of the countries that is part of the High Level Panel for a Sustainable Ocean Economy. Through this panel Canada has committed to an ocean action agenda – sustainably managing 100 per cent of our national waters by 2025, a daunting task! The United Nations Development Programme's Sustainable Development Goal 14 is called Life Below Water, and it sets out a vision via specific targets to "sustainably manage and protect marine and coastal ecosystems from pollution, as well as to address the impacts of ocean acidification."[88] If, after experiencing the richness of Botanical Beach's tidal pools, which are still considered a very healthy ecosystem, you have a desire to make a difference too, then consider the ideas outlined in the Engaging and Taking Action section below.

> The intertidal zone is such an interesting and dynamic place; these organisms are so incredibly well adapted to survive in such a harsh and extreme environment: Imagine living where you have to compete for space, suffer drastic temperature and salinity changes, and watch as your life-giving seawater ebbs and flows, all on a twice-daily basis! Knowing how to identify these organisms increases one's appreciation of this place, and understanding a bit of what they go through, day after day, is empowering, and even a little humbling.
>
> —**PHIL LAVOIE,** Bamfield Marine Sciences Centre, educator and naturalist

Engaging and Taking Action

The Surfrider Foundation is a North American non-profit with local projects that include beach cleanups, coastal preservation, ocean protection strategies, and ensuring beach access for all.

If you see a sick or dying sea star, please submit your citizen science observations to the UC Santa Cruz monitoring site, seastarwasting.org. Your

BY SEA, UCLUELET AND ITS AQUARIUM ARE ONLY 88 KM NORTHWEST OF BOTANICAL BEACH, BUT BY ROAD IT IS OVER 300 KM AND CLOSE TO FIVE HOURS' DRIVE! NONETHELESS, *IF YOU ARE WANTING A TRULY WEST COAST EXPERIENCE, THEN SPENDING TIME AT EACH LOCATION IS MORE THAN WORTHWHILE.* PHOTO JORDAN HAWKSWELL.

observations can help researchers track disease spread and understand the potential causes and consequences of sea star wasting.

On the west coast of Vancouver Island, the Ucluelet Aquarium has an array of public coastal conservation events each summer.

In Metro Vancouver, the Howe Sound Biosphere Region provides opportunities for the public to engage in coastal conservation. OceanWatch also provides a report and multiple ways to get involved in Howe Sound around its shorelines, estuaries, and the coastal species that inhabit the region.

You can make a difference at home too. For example, when we launder clothes made of fleece, polyester, and other man-made materials, they release microfibres that don't break down when they enter the environment. These fibres can often be found in the digestive systems of filter-feeders, such as clams and mussels, and in many species of fish. Consider the ways in which you could reduce the microfibres you release into the environment at surfrider.org. Start by washing your clothes, especially outdoor apparel, in cold water (see Metro Vancouver's Switch to Cold campaign for more information). Also at home, be thoughtful about the seafood you purchase. Learn about the environmental issues that surround cheap shrimp, prawns, and whitefish, or the concerns surrounding farmed salmon. Both Ocean Wise and Monterey Bay Aquarium's Seafood Watch guide are great sources of reliable information about sustainable seafood options, as are the certification programs that exist to inform consumers about sustainable harvesting of various ocean species. Look for Marine Stewardship Council (MSC) and Aquaculture Stewardship Council (ASC) labels on seafood sold at your local supermarket.

Minimizing Your Carbon Footprint

As Port Renfrew is at one end of the West Coast Trail, there are daily buses to access this area from both Victoria and Nanaimo, which run from May 1 to the end of September. So, if you don't mind sharing a ride with backpackers (they may smell better going in than coming out, but their stories are always better on the out!) consider the West Coast Trail Express. Port Renfrew is a 5 km walk from Botanical Beach.

Directions and Accessibility

Highway 14 is a spectacular coastal route that ends in Port Renfrew, 104 km from Victoria. Botanical Beach is another 5 km beyond Port Renfrew.

Accessibility to Botanical Beach is difficult. Though there is a wide gravel path to get to the beach area, the shoreline landscape is rough and has many obstacles such as driftwood. Contact Sooke–Port Renfrew Tourism for more information.

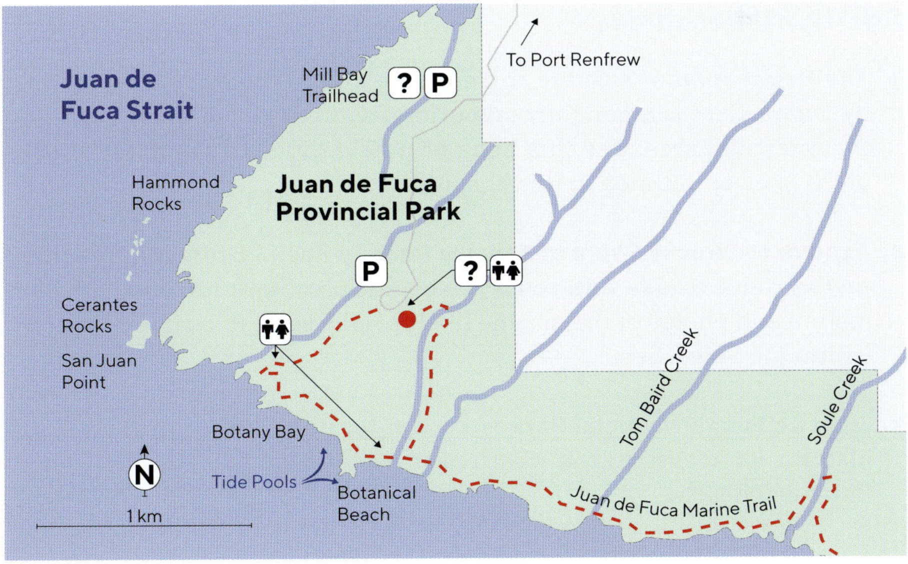

Other Locations to View Intertidal Species

1. Brady's Beach, next to Bamfield Marine Sciences Centre is another fantastic tidal area to explore.

2. Vancouver Aquarium has an amazing wet lab and display tanks of coastal BC species.

3. Visit Stanley Park at low tide, especially on the shores around Lumberman's Arch.

4. Ruckle Provincial Park on Salt Spring Island has tide pools that span the beach at low tide.

5. Rocky shorelines in and around West Vancouver such as Whytecliff Park, Lighthouse Park, or Caulfeild Cove are also great places to explore the intertidal zone; as are Bowen Island, Porteau Cove, and other rocky shorelines in Howe Sound.

Other Area Interests

Botanical Beach is at the northern end of the Juan de Fuca Marine Trail. Consider hiking pieces of this trail as there are several access points. China Beach is a favourite spot on the trail for many.

Additional Resources

1. ***The Beachcomber's Guide to Seashore Life in the Pacific Northwest*** by Duane Sept is among my favourite resources to use when exploring the intertidal area. Use this book as a field guide to see great images of the species outlined in this chapter.

2. ***Explore the Rocky Shore at Stanley Park*** by Sheila Byers is a free interpretive guide to take with you if you go tide pooling. It has detailed information and photos of the marine life typically seen at local rocky shores within the Salish Sea.

3. **Juan de Fuca Provincial Park** is known for its 47 km coastal trail, but also for its access to the west coast ecosystem, such as the area at Botanical Beach.

TIDE POOLING TYPICALLY INVOLVES WALKING OVER ROUGH AND SLIPPERY AREAS CLOSE TO THE OCEAN'S EDGE. *PAY CLOSE ATTENTION TO YOUR AND OTHERS' SAFETY WHEN EXPLORING IN REMOTE AREAS.*

4. **The Global Hub for Real-Time SDG Data** is a site that provides up to date data on all the Sustainable Development Goals (SDGs), including SDG 14, Life Below Water. On the site you can look more closely at datasets for this SDG related to topics such as bleaching of coral reefs and global fishing activity.

CLIMATE CHANGE AND INTERTIDAL SPECIES

The intertidal zone, which lies between the high and low tide marks on the shores of the world's oceans, is a sensitive marine ecosystem, and climate change is disturbing its basic characteristics, such as temperature and acidity. For example, many think the physiological stress on sea stars caused by warmer coastal waters has reduced their resilience in warding off the virus responsible for sea star wasting disease. Ocean acidification, a result of our warmer oceans absorbing more carbon dioxide, inhibits the larval development of shellfish, making it more difficult for them to build their calcium carbonate structures. Finally, warming oceans are affecting plankton growth, the foundational food of the ocean ecosystem, with cascading effects throughout the food web.

OFTEN, THE FIRST SIGN I SEE OF ORCAS IS THE MIST THAT RISES FROM THEIR EXHALE AT THE SURFACE. PHOTO TERRY BEREZAN.

Killer Whales & Marine Mammals

Shore-Based Viewing from East Point Lighthouse on Saturna Island

AT A GLANCE

WHAT: Killer Whales, also known as Orcas (*Orcinus orca*), Humpback Whales (*Megaptera novaeangliae*), Harbour Porpoises (*Phocoena phocoena*), Steller and California Sea Lions (*Eumetopias jubatus and Zalophus californianus*), North American River Otters (*Lontra canadensis*), and Harbour Seals (*Phoca vitulina*)
WHERE: East Point Regional Park (Tekteksen)
WHEN: June–September
TRADITIONAL TERRITORY: Á,LEṈENEȻ ȽTE (W̱SÁNEĆ), Tsawout, Tseycum, Tsarlip, sc̓əwaθenaʔɬ təməxʷ (Tsawwassen), Coast Salish, Stz'uminus, Hul'qumi'num Treaty Group
ACCESS: Easy

Have you ever been on a ferry when the captain calls, "Whales off the starboard bow!"? On BC Ferries this is sure to create a mass movement of passengers, an excited buzz of conversation, and many cold "sunshine breakfasts" as owners abandon them for a chance to see a marine mammal – once they figure out where starboard is, of course! That excitement is why half a million people actively watch marine mammals in the "Salish Sea" every year.[89] (The Salish Sea comprises the straits and sounds adjacent to the coastline of BC and Washington state, from Campbell River in the north, to Seattle in the south, and Port Renfrew in the west.) While most of these people intend to watch whales from boats, whale watching from boats is NOT the focus of this chapter. The reason for this is that the

Southern Resident Killer Whales (SRKWs) of the Salish Sea are *endangered*. There were only 74 of them left as of December 31, 2023, nearly a 25 per cent decline from the 98 Killer Whales (Orcas) that existed in 1995 (which was likely below their historical numbers), and their future is precarious. Their three main challenges are:

1. Boat vessel engine noise and disturbance. Killer Whales forage and socialize through communication and echolocation, and increasing boat traffic compromises this ability.

2. The population of Chinook Salmon, their main prey species, is collapsing in the Salish Sea.

3. Their food is contaminated with chemicals (toxins), primarily released in the past, but issues still exist today.

SHORE-BASED WHALE WATCHING PROVIDES AN INCREDIBLE EXPERIENCE WHILE VIRTUALLY ELIMINATING IMPACTS ON THE WHALES YOU ARE OBSERVING.
PHOTO LUCY QUAYLE.

PROMOTING SHORE-BASED WHALE WATCHING, THERE ARE WHALE TRAIL VIEWING SITES THROUGHOUT OCEANSIDE BC, AS WELL AS DOWN THE WEST COAST TO CALIFORNIA. SOURCE: WHALETRAIL.ORG.

Whale watching by boat can be a great experience, but in the Salish Sea and around the steadily decreasing population of Southern Resident Killer Whales, fewer boats lead to less acoustic noise, and this in turn is beneficial to Killer Whales, who rely on communication for successful hunting. Boat-based whale watching works in many areas, but it must not be carried on if it is detrimental to the species, so it really should be avoided as a means to view the SRKWs.

Therefore, my aim is to lead the reader to **land-based locations** from which they are most likely to spot a whale, thereby minimizing the impact of boats and boat noise on the whales themselves. According to the non-profit organization the Whale Trail, watching Orcas from shore is an *act of conservation*. The Whale Trail is a 2400 km stretch of over 100 shore-based whale viewing sites on the Pacific Coast between California and British Columbia. Land-based viewing for the SRKWs may or may not provide a whale experience; however, your impact will be less, and your experience will still be memorable, due to the wider ecosystem "experience" that will surround you in Gulf Islands National Park. This chapter is designed to help you enjoy a land-based experience.

Within a day's drive from Vancouver, among the best land-based viewing sites for Orcas and other marine mammals (including Humpback Whales, Minke Whales, North American River Otters, California Sea Lions, and

Harbour Seals), is the east end of Saturna Island at East Point Lighthouse, looking toward Boundary Pass and the Strait of Georgia. You can also watch for Harbour Porpoises, often seen foraging at the tideline during the changing of tides (check the tidal predictions for Narvaez Bay at tides.gc.ca). Saturna Island is the traditional territory of the W̱SÁNEĆ (pronounced wh-say-nuch) people of the Saanich Peninsula, comprising the Tsawout, Tsey-cum, and Tsarlip First Nations. East Point is known as "tekteksen" by the people who traditionally fished, hunted, and gathered around these islands. Tekteksen means "long nose" and refers to the shape of the land at East Point. Today, the WSÁNEĆ people have a 146-hectare reserve on Saturna at East Point.[90] Gulf Islands National Park comprises 43 per cent of Saturna Island.

It's a 15 km commute to East Point after arriving on BC Ferries into Lyall Harbour. The ecosystem "experience" begins even before arriving on the island, in Active Pass between Mayne and Galiano islands, or perhaps at the ferry transfer stop. But certainly by the time you set foot on Saturna and make your way to East Point, you'll notice that the Douglas Fir, Arbutus, and Garry Oak ecosystem – the trees, shrubs, ferns, and even birdsong – are noticeably different from those on the mainland. Once you arrive at East Point, a trail takes you up to the lighthouse or down to Tumbo Channel and Boiling Reef. Welcome to one of the stops (there are 21 others in BC) on the Whale Trail. Here, you can spend the remainder of your day scanning the horizon, picnicking, relaxing, and communing with nature. When you are patient and present, amazing things can happen! And if whales don't pass by, don't fret – you're at a remarkable vista, within a unique ecosystem, and there is sure to be other abundant wildlife to keep you entertained. Just being at East Point is a restorative act. You're in a community, side by side with the locals, sharing the pure joy and experience of seeing wildlife, just "doin' their thing!"

You will maximize your chance of seeing marine mammals by being prepared, so begin by learning to identify the marine mammals commonly seen here. Ocean Wise has printable whale resource guides for Cetaceans, Humpbacks of the North Coast, Pinnipeds, and Otters. The Center for Whale Research specializes in SRKWs in the Salish Sea. Marine mammals in this area include Humpback Whales, Killer Whales, Harbour Porpoises, Harbour Seals, North American River Otters, and possibly even an Elephant Seal.

If you arrive during the winter or a shoulder season, then both California and Steller Sea Lions may also be abundant. Low tides offer rock haul-outs for seals and sea lions, and foraging areas for deer, river otter, and mink, while high tides potentially bring whales closer to the shore. It's important to note, however, that whales' seasonal distribution changes with their food sources, and in recent years (2013 and 2016–18) the Southern Resident Killer Whales spent significantly less time in inland waters, likely due to the

PHONE APPS HELP REPORT WHERE WHALES ARE SEEN. THEY ARE USEFUL FOR BOTH SCIENTIFIC RESEARCH AND TO INFORM BOATS AND LARGER SHIPS TO ENSURE THEY SLOW DOWN AND GIVE THE WHALES SPACE AS THEY APPROACH. PHOTO LUCY QUAYLE.

lack of Chinook Salmon, while Bigg's Killer Whale (formerly known as Transient) and Humpback Whale sightings actually increased![91] In recent years, a mother Humpback Whale and calf have been seen from mid-April to the end of June, while Southern Residents were often seen throughout July and later in the fall (mid-October to early December). Bigg's Killer Whales are more prevalent in the spring when seal pups are born, and again are common in August, so plan accordingly.

The timing of whales is determined by both food abundance and their ability to *locate* this food. You can have salmon, but if the Orcas can't find them through echolocation and communication due to human noise, they aren't *available* to them! For Southern Resident Orcas, Chinook comprises 80 per cent of their diet,[92] so it's worth considering the human-caused reduction in the Fraser River Chinook Salmon run more seriously. A 2011 study looked closely at the amount of salmon required to feed all of the Southern Residents adequately (there were 87 at the time), and the researchers came up with 660 Chinook *per day*, which amounts to 12–23 per cent of the total Fraser River runs! Now there are only 74 SRKWs in existence, so

THESE TWO HUMPBACK WHALES, WHICH RETURN REGULARLY TO THE SATURNA ISLAND AREA, ARE NAMED HEATHER AND NEOWISE.
PHOTO PETRUS PIENAAR.

SEA LIONS FORM HAREMS OF ONE MALE AND MANY FEMALES. ***WATCH FOR THESE STELLER SEA LIONS AT BOILING REEF OFF EAST POINT IF YOU ARE VISITING IN THE WINTER.***
PHOTO SABINE DECAMP.

the USA recovery strategy goal to bring the population up to a stable 155 Orcas by 2030 would mean a 100 per cent increase in Chinook would be required to feed these Orcas.[93] The competition with humans for consumption of salmon, along with the existing impacts of climate change, helps us understand the enormity of the recovery challenge. The other whales you may spot feed on different species. Bigg's Killer Whales feed on marine mammals (including the seals when they are not hauled out on the nearby rocks, as well as porpoises, dolphins, and other small whales), and Humpbacks feed on krill and small fish.

Because of the suitability of Saturna's coastal features, this area is home to some of Canada's most intensive whale research, and the local organization for that is SIMRES, the Saturna Island Marine Research & Education Society. SIMRES is a community-based, non-profit organization that builds public awareness of marine life and their ecosystems through research, education, events, and an array of citizen-science undertakings. In planning a trip to East Point, I suggest reviewing what SIMRES has scheduled for the season, and plan to engage in a project, event, or one of their SeaTalks (see ideas below for Engaging and Taking Action). Much of the SIMRES research focuses specifically on the three pods of Southern Resident Killer Whales, called J, K, and L pods. These are the matriarchal social groups that

frequent the waters around Saturna, with J Pod being the most frequent. The pods swim along the kelp beds and past the sandstone cliffs of East Point between May and November as they travel from the outer coast of Vancouver Island to the Fraser River delta, one of their favourite feeding grounds. Most years the pods come together in a "**superpod**" in August or early September, for socializing and mating. Researchers have witnessed the whales appearing to have a "greeting ceremony" whereby one pod of whales will stretch out in a line to greet a new pod as it arrives![94] Male Orcas of these pods can live up to 70 years, and females can be reproductive until about 40, while living up to age 90! It's interesting to note the similarity to humans when considering how long females continue to live after menopause. The importance of matriarchal wisdom in leading a pod to productive foraging grounds is built into this species, and perhaps it's something we can learn from.

On summer weekends, volunteers open the picturesque historic Fog Alarm Building at East Point, where one can read the 1964 story of Moby

SHORE-BASED WILDLIFE OBSERVATION IS SPECTACULAR WHEN KILLER WHALES SHOW UP, BUT AT EAST POINT LIGHTHOUSE YOU CAN EXPECT TO VIEW MANY OTHER KINDS OF WILDLIFE, TOO. PHOTO JEANNE HYDE.

EAST POINT LIGHTHOUSE AND PARKS CANADA'S FOG ALARM BUILDING ARE ATTRACTIVE FROM THE WATER VIEW. *PERHAPS THIS IS THE VIEW WHALES WOULD HAVE OF YOU!*
PHOTO MAUREEN WELTON.

Doll, the first Killer Whale in the world to be captured (here off East Point) for a marine park. On Friday evenings in July and August, the Gulf Islands National Park interpreter organizes informal get-togethers on the grassy point adjacent to Boiling Reef. Often a local resident speaks about life on the island, and the gatherings may involve music, stories, and certainly laughs as everyone watches the sunset and moonrise (sometimes at the same time) from that vantage point.[95] There are so many opportunities to connect with both people and whales when you visit Saturna Island that you may want to spend multiple days engaging in, and learning more about, marine mammals of the Salish Sea. And don't forget, marine mammals aren't the only show in town: watch for Garter Snakes or Oystercatchers at Shell Beach, seabirds in the ocean, and Bald Eagles or Turkey Vultures overhead. These are all part of staying in place a little longer, so if that interests you, consider a Gulf Islands National Park camping experience at the walk-in-only, reservable Narvaez Bay campsite, or the boat-in-only Cabbage Island Campground and Marine Park, where vessels can moor overnight. There is also a private campground on Saturna called Arbutus Point.

I had the pleasure of spending an afternoon at the East Point Light house to interview Simon Fraser University master's student Lucy Quayle on one of her 106 days (845 hours!) of observation. From the point, Lucy's visual observations, and those compiled by volunteers of the Saturna Cetacean Sighting Network, were complemented by an underwater hydrophone acoustic network and GoPro time-lapse photography. Lucy, I learned, is

NARVAEZ BAY PROVIDES *RESERVABLE* WALK-IN OR CYCLE-IN CAMPSITES. *MIGHT THIS ENCOURAGE YOU TO MAKE YOUR TRIP TO SATURNA A CYCLING EXCURSION?*
PHOTO SABINE DECAMP.

partially supported by the Port of Vancouver, as they are interested to know whether container ships are complying with the voluntary slowdown from 18 to 11 knots, which it is hoped will reduce man-made disturbances for whales. It turns out that an impressive 60 per cent of ships were in compliance in the first year. Lucy also told me of seal pups that were birthed on the rocks in Tumbo Channel, and of Bigg's Killer Whales that "took an interest" in them. In Lucy's words, "I usually hear whales before I see them. It's their breathing that alerts me. Once you notice them, it can lead to some pretty significant events, such as the Bigg's that were swimming by the point, and then made a hard turn in the direction of Boiling Reef to hunt the seal pups hauled out on the rocks. Just incredible." It's a memory she won't forget! I learned, too, that one of Lucy's biggest concerns is the **bioaccumulation** of pollution and toxins up the food chain, building up in a whale's body, leading to poorer health. That said, Lucy is also hopeful, and spoke about how encouraged she is to see the huge rebound of Humpback Whale numbers on the Pacific Coast. Her research work is "old school" observation, which is critically important to understanding these whales.

Lucy's data complements the high-tech research methods, such as "Passive Acoustic Monitoring" stations, which are interspersed throughout the Salish Sea and monitor underwater sounds while building "Deep-Learning Sound Neural Nets." These allow for real-time detection and differentiating between Killer Whale "ecotypes" (Bigg's or Southern Resident), classification of

THE MARINE EDUCATION AND RESEARCH SOCIETY HAS AN IMPORTANT EDUCATION CAMPAIGN FOR BOATERS: SEE A BLOW? GO SLOW!

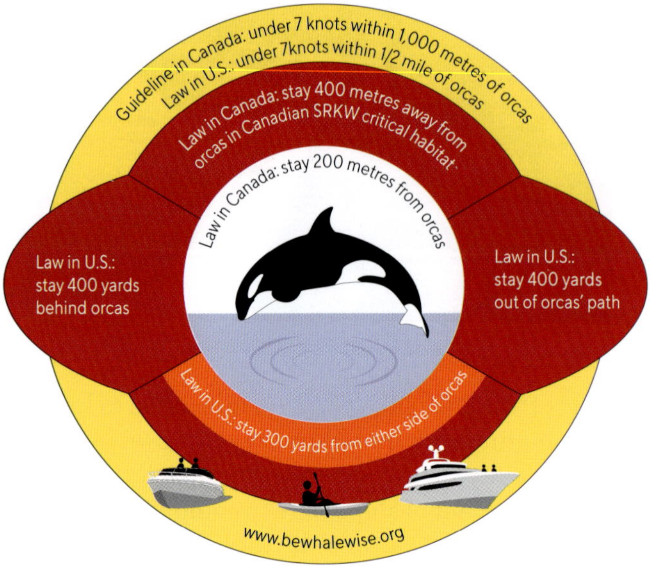

THIS WHALEWISE GRAPHIC, OUTLINING THE NEED TO KEEP A MINIMUM 400 M DISTANCE FROM ALL KILLER WHALES, GIVES AN IMPORTANT MESSAGE TO REMEMBER WHILE BOATING.

SRKW pods, and predictions about which direction they are moving in (to allow for the ability to send out alerts to ships).[96] In addition, the Canadian Coast Guard has set up a marine mammal desk with modern technologies to "report whale sightings in real time and advise vessel traffic by providing enhanced awareness of the activities of endangered Southern Resident Killer Whales and other cetaceans" (cetaceans are the family that includes whales, dolphins, and porpoises).[97] From the East Point Lighthouse, look southeast toward Boiling Reef and Patos Island in the USA, and southwest to Haro Strait where the container ships coming from overseas pass through the Strait of Juan de Fuca. There is an Interim Sanctuary ("no-go") Zone for all recreational and commercial vessels around Boiling Reef, and various regulations help to slow down pleasure boats and large ships in order to minimize both noise and ship strikes on whales, which can injure or kill these amazing creatures. Between 2004 and 2013, nine Southern Resident Killer Whales died from blunt trauma from ship or propeller strikes![98] Collisions (and fishing-debris entanglement) are human impacts that can be mitigated. If you want to get a sense of the marine traffic overlap between the shipping lanes of the Strait of Juan de Fuca and Georgia Strait and the critical habitat of the SRKW, have a look at the live ship traffic map on marinetraffic.com and pair what you see with HappyWhale.com, a site that links whale sightings to individual whales and documents their travel routes. Technology like this helps us build personal relationships with the ocean and its wildlife, and when we do I think we become more invested stewards of the marine environment.[99] Regulations state that *all* boats, including commercial tour operators and recreational boaters, must stay a minimum of 400 metres from the whales.

How much do you know about the **Sustainable Development Goals** (SDGs)? There are 17 of them and they include 169 targets. They are meant to be the guiding goals for the world from 2015 to 2030. SDG 4 is *Quality Education,* and one key is SDG Target 4.7. The target aims "by 2030 to ensure all learners acquire the knowledge and skills needed to understand and promote sustainable development...through sustainable development education...and sustainable lifestyles...and appreciation...of a culture's contribution to sustainable development."[100] This chapter is written to help the reader, the "learner," discover Orcas and marine mammals through sustainable practices and to engage in that learning through opportunities to support our wild whales. These range from knowing how you can report a whale sighting, to understanding Whale-wise guidelines and sanctuary zones, to attending a SIMRES event, or volunteering in habitat restoration. *The SDGs provide a blueprint for a sustainable future,* and my hope is that this chapter connects directly with Target 4.7 to help you learn what you can do today in reference to the global goals.

There's a Parks Canada sign at Monarch Head on Saturna Island in Gulf Islands National Park that reads:

> **"Standing up here on Monarch Head, you may feel like you are only a spectator. But have you eaten seafood from the waters below? Have you purchased anything those freighters are carrying? Have you flushed a toilet in this watershed, used a vehicle in this airshed? Everyday choices make a difference. What are you doing to restore and protect the health of our shared home?"**

I believe it's possible to "protect the health of our shared home," but only with many advocates. Let's all play our part, roll up our sleeves, and add our voices and efforts to meet this challenge.

> Among other animals, whales and dolphins seem truly curious about us. Their willingness to interact – their playfulness – has proved costly for them. How do we cross that bridge, without putting them further at risk? The world is shifting, and we must change our model of it if we want it to flourish and go on. Are the animals here just to entertain us, or meet some other need? We must acknowledge their right to exist for their own sake. Our best role is as guardians, ensuring them safe passage, and a chance to go on.
>
> —**DONNA SANDSTROM,** founder/director of The Whale Trail

Engaging and Taking Action

Ideas to engage include volunteering with SIMRES as they partner with the BC Cetacean Sightings Network to survey Harbour Porpoises. Begin by downloading the WhaleReport app, or if you have already made a sighting, by calling 1.866.I.SAW.ONE. SIMRES also partners with the SeaChange Society to plant and restore eelgrass beds. Eelgrass is a plant that forms the base of a highly productive marine food web, providing shelter for young fish, spawning surface for herring, and food for migratory birds. Eelgrass planting by volunteers often takes place during the lowest tides of August. Check the SeaChange website for opportunities to volunteer for restoration work.

SIMRES also offers an array of educational opportunities, such as their Intertidal Safari in partnership with the Marine Life Sanctuaries Society, as well as their SeaTalks for both adults and children. Finally, plan to learn about the BC Coast Hydrophone Network, of which SIMRES is a part, with hydrophones deployed in Boundary Pass and off Monarch Head. SIMRES is involved in a number of research collaborations relating to noise and human disturbance of the endangered SRKW.[101]

The Mayne Island Conservancy has extended a mapping project for bull kelp to Saturna Island. This is especially important work as kelp forests are the critical habitat (like old-growth forests on land) for marine mammals, fish, invertebrates, and seabirds.

Southern Gulf Islands Whale Sighting Network is a citizen science–based initiative providing local "eyes on the water" to understand whale frequency, timing, and species diversity. In addition, they have a library of cetacean photos, educate about conservation, and record violations in the sanctuary zone. When you're on Saturna, consider connecting with this group to learn the latest news!

Finally, at orca-live.net you can listen live on your phone to whale sounds from hydrophones farther up the coast on Hanson Island in Johnstone Strait (also part of the BC Coast-Wide Hydrophone Network), or watch for a real-time forecasting system for Southern Resident Killer Whales of the Salish Sea, currently in the works at Simon Fraser University.

Minimizing Your Carbon Footprint

Consider travelling on BC Ferries as a foot passenger, or with a bike instead of a car. Saturna is a beautiful island to cycle around. As with other Gulf Islands, there are quite a few hills, so an e-bike would also be a good option.

If you are a novice cyclist, be sure to let the ferry car traffic depart the terminal before you start your ride onto the island, just to minimize the number of vehicles that pass you on the road. If you are walking onto the island, consider taking the Saturna Lions Club shuttle. It's also possible that local accommodation may have options for you.

Directions and Accessibility

From Vancouver, take Highway 99 to Highway 17 and the Tsawwassen Ferry Terminal. If driving, ensure you have a reservation set up for the Gulf Islands Ferry to Saturna Island. You may have to transfer ferries at another Gulf Island to reach Saturna. Upon arrival at Lyall Harbour, take East Point Road to Tumbo Channel Road, which leads to East Point Regional Park. For boats travelling to Saturna, be sure to learn where Transport Canada has set up the Interim Sanctuary Zones off Saturna and Pender islands.

Contact the Saturna Island Tourism Association about accessibility. The lighthouse area and viewpoints are a mix of gravel and dirt trails. The association would be able to give more thorough directions to potential viewing locations on the water.

Other Locations to View Killer Whales

1. Lime Kiln Point on San Juan Island, Washington, includes an interpretive centre with hands-on exhibits and displays about Killer Whales. On the Saturday closest to the summer solstice each year, the City Cantabile Choir performs a free concert for the whales at an event called OrcaSing. Learn more at Friends of Lime Kiln Society.

2. Sheringham Point, just west of French Creek Provincial Park on Vancouver Island is a great location to spot whales or other marine mammals.

3. Otter Point, just east of Sooke on Vancouver Island is a potentially good shore-based location to view whales.

4. Beechey Head in East Sooke Regional Park, Vancouver Island is outlined in the Turkey Vulture chapter, and it would also be a good location for whale or other marine mammal viewing.

5. Gowlland Point or Oak Bluffs on South Pender Island are both stunning sites, and if you see whales from there, it would be a bonus!

HARBOUR SEALS HAVE EARS – AND NOSTRILS – THAT AUTOMATICALLY CLOSE WHEN UNDERWATER. THEY ALSO HAVE BIG EYES, ALLOWING THEM TO SEE IN THE LOW LIGHT OF THEIR DEEPER DIVES. THESE ARE OFTEN THE MOST COMMON MARINE MAMMALS PEOPLE SEE WHEN VISITING THE GULF ISLANDS. *BEGIN LOOKING FOR THEM IN ACTIVE PASS, WHILE YOU'RE ON THE BC FERRY.* PHOTO MICHAEL SCHMIDT PHOTOGRAPHY.

Other Area Interests

This is a good time to address, and speak briefly to, other marine mammals that may be seen while viewing whales from shore.

1. Steller Sea Lions overwinter from mid-October to March on the rocks at the Belle Chain Islets, viewable from Winter Cove at Boat Pass on Saturna Island. Both Steller and California Sea Lions can be seen, often vying for territory, on Boiling Reef at East Point. Both the Southern Resident and Bigg's Killer Whales are present in the winter too, with the latter watching for territorial disputes between sea lions that result in one being pushed off the rocks and into the buffet zone of the Bigg's! One of the beauties of visiting the Gulf Islands in winter is the opportunity to see an array of species, including bird life, that isn't around in the summer.

2. Harbour Seals haul out at Boiling Reef when low tide exposes the rocks. Known to local sea kayak guides as "rock sausages," at up to 1.8 m (6 ft.)

RIVER OTTERS CAN OFTEN BE SEEN FORAGING IN INTERTIDAL AREAS WHEN THE TIDE IS OUT. KEEP AN EYE OUT FOR THIS ENTERTAINING ANIMAL THAT IS COMFORTABLE BOTH ON LAND AND IN THE OCEAN. PHOTO CHRISTINE GAIO.

and 160 kg (350 lbs.), these are among the most common marine mammals spotted on the BC coast. Laden with fat, the seal pups are a staple in the diet of Bigg's Killer Whales. It's possible that if you see Orcas from East Point Lighthouse, they won't be Southern Resident Killer Whales but rather Bigg's passing by to see if young Harbour Seals may be on the menu.

3. North American River Otters are one of two otter species in BC that spend equal parts of their time on land and in the water. The other BC otter is the larger Sea Otter, which lives almost completely in the ocean. River Otters protect themselves from cold Pacific waters in a very different way than Harbour Seals. Whereas the seals use fat for insulation, the otters have thick, water-repellent fur. These members of the weasel family are commonly seen feeding off the coasts all around Saturna. It is great fun to see this species with fully webbed toes jump into the water and swim effortlessly. Although Harbour Seals can be spotted at just about any time of day, North American River Otters are most active at dawn, dusk, and throughout the night.

4. Consider a kayak rental or tour with a local sea kayaking company to explore safely around Samuel Island, Mayne Island, and the Belle Chain

THE DORSAL SADDLE PATCH PATTERN AND NICKS AND SCRATCHES ARE USED BY RESEARCHERS TO IDENTIFY INDIVIDUAL KILLER WHALES. *LOOK FOR THE PRINTABLE IDENTIFICATION GUIDES ON THE OCEAN WISE WEBSITE.* PHOTO JEANNE HYDE.

Islets, or to stay overnight on Cabbage Island. On a clear day, a walk along Brown Ridge atop Mount Warburton Pike, the highest point on Saturna Island, will provide stunning views, and you may even come across some of the local *wild goats*. Visit Monarch Head and Narvaez Bay for a slightly different view of Boundary Pass. There is also a weekly Saturday market (10–1) that runs between the Victoria Day and Labour Day weekends.

Additional Resources

1. **The Whale Trail** is a non-profit that promotes shore-based whale watching.

2. **Ocean Wise** is a good place to begin to understand practices around viewing Orcas by boat through their Whalewise regulations.

3. **Happywhale** is a non-profit that tracks individual whales throughout our world's oceans and inspires kinship between humans and marine life through whale-based citizen science.

4. **Georgia Strait Alliance** is a conservation organization dedicated to keeping the public informed about the latest conservation developments related to the SRKWs. Consider joining their Orca Action Team.

5. **Cetus Society's Straitwatch** is a conservation organization that actively intercedes to protect whales from the harm caused by vessel traffic.

6. **Raincoast Conservation Foundation** – Understand the potential impacts on Killer Whales with the Port of Vancouver's Terminal 2 Expansion.

7. **The Port of Vancouver ECHO Program** (Enhancing Cetacean Habitat and Observation) is an important addition to changing how shipping containers move through the Strait of Juan de Fuca.

8. ***Hakai Magazine's* "On the Trail of Whales"** – *Hakai* explores science, society, and the environment from a coastal perspective. Their stories are about the coast – past, present, and future.

9. ***Salish Sea Wild*** is a series of videos about the Salish Sea, created by the SeaDoc Society, an organization that conducts and sponsors scientific research in the Salish Sea and translates that science into action.

10. **DFO Marine Mammal Incident Reporting Hotline** – For emergency marine mammal incidents you observe such as distressed or deceased marine mammals (1-866-I SAW ONE).

CLIMATE CHANGE AND KILLER WHALES

The Southern Resident Killer Whales are mainly salmon eaters, and climate change is affecting salmon populations, so the SRKWs are particularly susceptible to impacts on their primary energy source. Harbour Seal populations are increasing and they are a key source of food for Bigg's Killer Whales, who may benefit in the short term, but our overall understanding of the ways climate change is impacting ocean ecosystems still has a long way to go.

AUTUMN

ALTHOUGH SOCKEYE SALMON RUNS HAVE DECREASED SUBSTANTIALLY, THERE ARE STILL AREAS WHERE ONE CAN SEE THIS SPECTACULAR AUTUMN EVENT. SOME RUNS, SUCH AS THE ADAMS RIVER RUN, PEAK EVERY FOUR YEARS. PHOTO ERIC GROSS.

INTRODUCTION TO AUTUMN

Back to School, in More Ways Than One

BY JOHANNA WAGSTAFFE

Meteorologist and science reporter for CBC News Network and CBC Vancouver. Covers national and international natural disasters and breaking science news. Host of three CBC award-winning podcasts: *Fault Lines, 2050: Degrees of Change,* and *Uncover: Bomb on Board*. Author of children's science books *Fault Lines: Understanding the Power of Earthquakes,* and *Little Cloud: The Science of a Hurricane.*

Fall feels like a time for fresh beginnings. Although *nature* will say beginnings happen in spring, I just can't shake those back-to-school vibes.

As the sun sets a little earlier on those late summer days, I feel the familiar twinge of the dog days winding down. It's always bittersweet to see the first changes of colour in the deciduous trees when, taking their cues from the cool overnight temperatures and diminishing length of daylight, the trees begin their chlorophyll shut-down process. As a broadcast meteorologist, I am taking cues from viewers and listeners wanting to know how vibrant the leaves' colours will be this year, and what kind of winter lies ahead. I begin the windup process after summer programming. I get ready for the fall news cycle, which brings new show launches and fresh eyes from all the bosses. I remind myself to brush up on fall storm forecasting and first-frost averages. While the official equinox doesn't happen until late September, *my* fall (going by the "meteorological seasons," as we do for easier climatological forecasting and research) runs from September 1st through to November 30th. I may not be buying new school pencils anymore, but, even well into adulthood, autumn always means a *fresh beginning* for me.

For the first time in a long time, though, I do truly feel like a student again – because I've spent the past two autumns as a mother. Together, my young son and I have now seen two full fall seasons of change. And although he was a spring baby, his first fall was the moment I really saw him open his eyes to nature. By his second year, he was full-on running, climbing, and chatting with all the world around him.

We spend almost every morning of the year outside, rain or shine, walking in the Lynn Valley woods with my dad and my dog. By fall, we're racing against the changing angle of the sun to pick "just one more" handful of blackberries. And then the true signs of autumn set in. The air is crisp, the migratory birds begin taking flight, the streams get fuller and faster, and the leaves begin to crunch underfoot. At first, I feel overwhelmed by the new species we are noticing, on the ground and in the sky. I remember the first time my son hears (and then spots!) a woodpecker with delight. I tell him its common name, and he blinks up at me wanting more. So I turn to the ornithological investigators online and, through social media, learn that *she's* a Red-breasted Sapsucker. He seems pleased with that answer.

As I frantically google every new bird's song we hear and mushroom's silhouette we spot, I realize what a joy it is to be learning as my son is learning, sharing in his discovery and wonder. I discover that British Columbia rests below a major migratory thoroughfare that stretches 15,000 km between Alaska and Patagonia.[102] I'm surprised to find that there are more than 50 types of mushroom species in the Pacific Northwest alone,[103] and that my son is an expert fungi forager. (We decide it's best not to touch any of them until I get better at identification!) We learn about the best rivers to visit to see the salmon run – and that it's probably wise to leave the dog at home for those adventures.

I think of my grandmother taking my sisters and me through these very same woods when we were children, and how she must have felt the same pride that I do now, being both a teacher and a student in these surroundings. She taught us about the different kinds of slugs, the best trails for finding the ripest berries, and that when you get stung by the nettles, relief can be found in the curing, nearby Dock Leaf.

But there's something else that happened in my son's second fall. In mid-September 2020, millions of British Columbians were trapped indoors, in the middle of the pandemic, when being outside and in touch with nature couldn't have meant more. Smoke from an unprecedented wildfire season across the USA west coast was carried up into BC, where it sat for ten days. Unusually strong, easterly outflow winds sent thick plumes of smoke from Washington, Oregon, and northern California out over the Pacific Ocean, where they curved toward the north before turning back east, invading the

A TURKEY VULTURE IN FLIGHT OFTEN HOLDS ITS WINGS IN A V SHAPE. PHOTO TERRY CARR

entire southern coast of British Columbia. The Pacific's cold September surface waters created temperature inversions that helped to trap the smoke in, and the longer nights of September helped to keep those inversions around longer. There was no relief overnight. As a result, air quality levels for much of southern BC surpassed the worst multi-day averages ever recorded, even though 2017 and 2018 had already established recent new highs. Total hours of restricted visibility over the ten-day event were almost double those of 2018![104] Air quality meteorologists said the smoke event differed from previous years for two significant reasons: the smoke arrived later than usual; and it came from fires that were not on our own soil.

My son and I stayed indoors, unable to see even a few metres beyond our window at times. Our plans to explore the nearby wilds were suddenly put on hold – indefinitely.

Climate change will impact all of our seasons in different ways. According to the Pacific Climate Impacts Consortium, British Columbia has already warmed an average of 1.4°C between 1900 and 2013. While the warming is most pronounced in our north, BC's southern coastal regions have warmed 0.8°C since the Industrial Revolution, roughly equivalent to the global average rate. We've also seen most of our warming in the winter and in our overnight temperatures. Even in a middle-of-the-road scenario, in which we manage to decrease future greenhouse gas emissions by half, the average annual temperature in BC would still increase more than another full degree by 2050![105] That is a profound escalation that will have untold negative impacts, not just for our weather patterns and seasons but for the plants and animals that call this province home too.

The trends we are already seeing as a result of climate change will continue. Summer droughts will extend, wildfire season will begin earlier and end later, the first frost will be delayed, and our rivers and streams will be both warmer and lower. I have witnessed these changes as a meteorologist: the shifting baselines and trends, the more frequent extreme events. I have reported on the big ones: the record fire seasons, the floods, and the deadly heat waves. We've all suffered the inconvenience of the necessary water restrictions. Global warming will continue to bring about big, inescapable change. And it will do so with greater frequency and severity over time.

But it's the little changes to our shared home that got me thinking this past fall. And I learned about a new word that strikes a chord with me: ***solastalgia***. It refers to a melancholy brought on by changes to the natural world, changes becoming evident all around us.[106] Because maybe in 30 years many birds won't migrate anymore; the berries could be out well into October; and the mushrooms might lean more heavily toward exotic southern species. Changes will come slowly over time, season by season.

I am only a couple of years into the greatest job I'll ever have – being my son's mother. But every time we get set to "go back to school," I will remember the markers and the memories shared with him. And I will hope that when it's his turn to pass on his knowledge, he will remember the song of the woodpecker preparing for winter. I hope he won't feel sad if the Red-breasted Sapsucker no longer calls southern BC home, but rather that he'll be happy that, when we had the opportunity, we stopped the worst of climate change from coming to pass. That we can still share our fondest memories and our most important experiences. That we learned from Indigenous communities how to care for our home. That we adapted to the new world, a world that will still be full of berries and birds. That is my greatest hope. So let our birdwatching not only bring us joy but also serve as a record of the extraordinary natural world around us, a world we have the privilege of bearing witness to, a world to be cherished and passed on – together.

MANY TURKEY VULTURES MAY BE SEEN FLYING OVERHEAD AT EAST SOOKE REGIONAL PARK IN MID-SEPTEMBER IF THE UPDRAFT CONDITIONS ARE RIGHT. PHOTO. ERIC GROSS

Turkey Vultures & Raptors

at East Sooke Regional Park, Vancouver Island

AT A GLANCE

WHAT: Turkey Vultures (*Cathartes aura*) and Raptors kettling in the sky
WHERE: East Sooke Regional Park
WHEN: Second week of September to first week of October (prime viewing in third week of September if sunny and warm)
TRADITIONAL TERRITORY: Á,LEṈENEȻ ȽTE (W̱SÁNEĆ), Coast Salish, S'Klallam, Te'mexw Treaty Association
ACCESS: Easy to moderate

Look at a map of Vancouver Island and you'll see that its geography conspires to concentrate numbers of migrating birds at the southern tip of the island, ensuring they fly the shortest crossing of the Strait of Juan de Fuca on their way to the Olympic Peninsula and beyond. This location occurs 40 km west of Victoria in East Sooke Regional Park, and many birds, including Turkey Vultures and several species of raptors (mostly eagles, hawks, and falcons), soar on the warm **thermals** that intensify off the rocky bluffs on warm days between the second week of September and first week of October. The very best scenarios for vulture viewing occur when a warm day follows several cool or rainy ones, or when there is cool fog over the strait, keeping the birds on the island. After a spring and summer on the island, the birds arrive at this staging area and are faced with a watery barrier before continuing their southern journey. They are part of a wider western population of over a million birds that migrate south to the USA and Central America, and some as far as Venezuela, Colombia, and Ecuador for the winter.[107] In fact, Turkey Vultures are the widest-ranging vulture in the New World, found throughout southern BC in open country, especially when that landscape is within a few miles of rocky or wooded areas.[108]

AN OVERHEAD FLIGHT OF A TURKEY VULTURE SHOWS BOTH ITS 1.7 M WINGSPAN AND THE TWO HUES OF ITS FLIGHT FEATHERS. PHOTO ROB ALEXANDER.

Turkey Vultures are not strong fliers and therefore take advantage of currents of rising air to gain height. When foraging, they do this to get to a higher vantage point, but at East Sooke Regional Park the height is required to ensure they are successful on their 20 km water crossing. To find the best rising currents, they require teammates as "thermal detectors," and a group is necessary for that. Therefore, in late September, you can see upwards of 400 Turkey Vultures and raptors "**kettling**," that is, taking advantage of rising thermal air masses and using them to swirl and spiral higher. The warmer the day, the more likely they will cross, so monitoring Sooke's weather forecast prior to your trip is helpful. The Turkey Vultures cross the strait in single file, en route to the Olympic Peninsula. Use binoculars to distinguish the juveniles that hatched in May, as they'll still have grey heads. They migrate south as a group, and travel long distances without feeding. Don't be surprised, however, to see the Victoria Christmas bird count numbers include Turkey Vultures overwintering on the island, either because they didn't make the crossing while the land and water were still warm enough to create strong thermals, or because they decided the South Island was a fine place to stay for the winter.

East Sooke Regional Park is large (50 km of trails and 31 km in diameter), and it would take a full day to hike in one direction across it; therefore, choose your trails and vulture viewing areas wisely. A popular place to begin is the Aylard Farm parking area, where there is decent viewing at the farm, out on the coastal bluffs, or, if the fog is rolling in, up higher at one of

VULTURES RISE ON WARM AIR CURRENTS AT SIMILAR TIMES, LEADING TO GROUPS IN THE SKY. THIS GROUP BEHAVIOUR IS KNOWN AS KETTLING, AND IT PROVIDES VULTURES WITH "THERMAL DETECTOR" TEAMMATES TO HELP THEM FIND THE RISING CURRENTS.
PHOTO ERIC GROSS.

the viewpoints. My favourite route is the undulating Coast Trail to Squeaky Point, past the petroglyphs at Alldridge Point, and on to the most southerly point of the park at Beechey Head (6.1 km if you return as a loop). A high-elevation location in the park is the challenging 12.3 km Mount Maguire loop, accessed from the Pike Road parking area. Be prepared for a four-hour, 630 m elevation gain to get to the 879 m peak. Whatever else you do, make sure you find a place to lie back on the rocks, observe the sky, and identify whatever you can.

Male and female Turkey Vultures are virtually identical, large blackish-brown scavenger birds with a 1.7 m, or more, wingspan. These vultures may be mistaken for eagles, but they aren't difficult to identify once you know what to look for. A Turkey Vulture is a soaring bird, so it's rare to see one beating its wings. A Turkey Vulture's wings are tilted above their backs in a V shape, allowing it to rock or "wobble" back and forth when soaring. Its two-toned, silvery flight feathers look like long "fingers," and its wing-fronts are a darker black. Its tail is also lightly coloured and extends past

TURKEY VULTURES RARELY FLAP THEIR WINGS. INSTEAD, THEY SOAR WITH A DISTINCTIVE ROCKING MOTION, THEIR WINGS TILTED ABOVE THEIR BACKS TO FORM A V SHAPE.
PHOTO TERRY CARR.

its feet. The red head (featherless and reminiscent of a wild turkey) of the vulture is small and won't always be noticeable when soaring above, but binoculars will help. In contrast, Bald Eagles flap and glide with dark, single-toned wings that are broad and flat. Once you've learned to identify Turkey Vultures, you may be drawn to the greater challenge of identifying the other raptors often present in the rising air masses, including Bald Eagles, Red-tailed Hawks, Peregrine Falcons, Merlins, American Kestrels, Cooper's and Sharp-shinned Hawks, Northern Harriers, and Ospreys.

Raptor identification is an important birding skill, and there are several things you can do to accelerate your learning curve. Start with the Merlin Bird ID mobile app. Download the Western Canada bird pack and attempt to use it at the park. Read up on and review the raptors common in BC at Allaboutbirds.org. Consider an online course or two at the Cornell Lab Bird Academy or locate a *Hawks in Flight* book, or an online web page such as Audubon's Identify Raptors in Flight. However, these are all solo ways of learning and, although they have a place, you can't beat learning from the depth of knowledge and experience of others. So also consider field trips led by the Victoria Natural History Society or the community workshops provided through the Rocky Point Bird Observatory. Increasing your bird

A PHONE APP LIKE CORNELL LAB'S MERLIN IS VERY USEFUL FOR BOTH VISUAL AND AUDIO IDENTIFICATION OF BIRDS. INSTEAD OF YOU HAVING TO SCROLL THROUGH BIRDS TO FIND WHAT YOU'VE OBSERVED, IT ASKS YOU FIVE SIMPLE QUESTIONS. THE ANSWERS LEAD TO SUGGESTIONS ON WHAT YOU MAY BE SEEING.
SOURCE: THE CORNELL LAB.™

identification skills is a lifelong endeavour. I'd say for me it's among the most important elements that helped me connect more deeply to the natural world and, in turn, directed both my education and career path.

Turkey Vultures, like bats and snakes, or large predators like bears and wolves, are often maligned, but that may just be a reflection of a lack of education. As scavengers, vultures are associated with death, and there can be historic prejudice stemming from stories that arose in our agrarian past. Anthropomorphizing Turkey Vultures can elicit feelings of disgust when seeing them feed on carrion (decaying dead animals), or when they defecate on their legs, which is actually an effective thermoregulation adaptation used to cool themselves on hot days! People's negative opinions are often largely based on ignorance, from not fully understanding the critical role maligned species play or the ecological niche they fill in nature. Turkey Vultures, for instance, have evolved potent stomach acids to protect themselves against the bacteria and pathogens that form on a carcass. By stripping dead animals to the bone, vultures keep these microbes from infecting other species, including our livestock and us. To locate carrion, vultures have a well-developed sense of smell (rare in a bird) and are known to have the largest olfactory bulb of any of the world's birds.[109] They can detect the gas that rises off carrion from up to 1.5 km away. They prefer to feed on smaller and freshly dead animals, as they're unable to break through the carcasses of large mammals with their beaks. That means Turkey Vultures have to wait for ravens or eagles to break the skin of a dead deer or the carcass of a washed-up seal. Unlike those species, however, they can't remove meat to eat elsewhere, so you tend to see them eating from a carcass where it died, which is sometimes roadkill in the middle of a rural road – not the safest place for any species!

Though Turkey Vultures and their young are preyed upon by predators like Bald Eagles, Ravens, Racoons, and Rats, the greatest impacts on this species stem from humans. As with other large birds, electrocution on power lines and collisions with vehicles on roads are serious threats. People shooting them directly also still happens, but this has decreased as regulations and policy have moved in the direction of protecting them. One issue, however, that is still a prevalent threat for this species is bioaccumulation. **Bioaccumulation** is the gradual accumulation of a toxic substance, such as pesticides or chemicals, in an organism.[110] If a species accumulates toxins faster than it can eliminate them, they risk chronic poisoning. An example of this in birds is the buildup of the historic pesticide **DDT** (Dichlorodiphenyltrichloroethane), now banned in North America but not fully eliminated from the environment, nor from all countries. Pesticides accumulate in insects that feed on agricultural crops sprayed with DDT. Insect-eating birds forage for these bugs, which have become pesticide-laden, and the poison within the insects accumulates in the birds'

THE FEATHERLESS RED HEAD OF A TURKEY VULTURE IS REMINISCENT OF A WILD TURKEY, HENCE ITS NAME. AS CARRION EATERS, VULTURES MAY BE BETTER ADAPTED WITHOUT THE HEAD FEATHERS THAT ADORN MOST OTHER BIRDS. PHOTO ROB ALEXANDER.

tissues. Larger birds prey on the insect-eating birds, so the concentration of poison increases further up the food chain. The result is that the top predatory species such as Bald Eagles and Peregrine Falcons, as well as scavenging species such as vultures, become less able to form eggshells hard enough to protect their young, leading to nest failure. Bioaccumulation is also a significant issue with Orcas (see chapter on Killer Whales) from long since discarded PCBs. (Polychlorinated Biphenyls are extremely persistent chemicals that last in the environment for many years. They were used industrially for four decades prior to being banned in North America in 1977 and cellular tissue of animals today still show their presence.) When blubber is sampled from an Orca, it is still found to contain extremely high amounts of this toxin, likely from the prey it eats *and* from the prey its prey eats![111]

For the Turkey Vulture, the major bioaccumulation issue is lead poisoning from the ingestion of animals killed and left by hunters using lead shot, or dead fish that have ingested lead jigs or sinkers. Even small amounts of lead can sicken or kill vultures, because when the lead reaches the acidic environment of the stomach, it dissolves and is absorbed into body tissues. As the lead reaches toxic levels in the tissues, muscle paralysis and other complications result in death. And since a vulture's entire diet is composed of dead animals, they are more frequently exposed to lead bullet hazards than are most wildlife.

Climate change is another factor affecting most bird species, but for the Turkey Vulture it seems to be a neutral factor or maybe even somewhat positive, as it has the potential to increase their northern range. Audubon has set up an information-rich website called Survival by Degrees: 389 Species on the Brink, and it allows the public to use their database of 140 million observations to look at the range of 604 bird species in North America, and project how temperature increases of 1.5°C, 2.0°C, and 3.0°C will affect these species. As the site states, "The results are clear: Birds will be forced to relocate to find favourable homes. And they may not survive."[112] That said, if you look at Turkey Vultures through this climate change lens, you will see they are more resilient than most birds, as they are relatively adaptable in terms of both their diet and their nest sites.

The current population of Turkey Vultures in BC appears to be stable or increasing. An important way to help mitigate impacts is to collect data on a species by tracking them, which leads to an increase of our understanding about their habitat needs and their population dynamics. That often means attaching tracking tags, such as the **Icarus satellite tag**, especially useful for long-distance migrants like the Turkey Vulture. The tag is a small clip, weighing less than 3 per cent of their body weight. A mobile phone user can download the Animal Tracker app and literally follow tagged species throughout the world! For any migrating bird like the vulture, there are four key areas we should be aware of, because what we do locally is only part of their story. Satellite tags help scientists monitor the species in (1) breeding grounds, (2) wintering grounds, (3) northern migration routes, and (4) southern migration routes. Tracking birds throughout their travels is critical to our understanding, and therefore protection, of them, and the technology to do so is becoming more miniaturized, less expensive, and able to transmit more and more data.

A study published by conservation biologists at the Cornell Lab of Ornithology and the American Bird Conservancy announced that by 2019 North America had lost nearly three billion birds in the past *half-century*. Ann Nightingale, from the Rocky Point Bird Observatory (see Engaging and Taking Action below), loves birds as much as anyone I know, but she told me that it's very hard to be optimistic about their long-term survival, unless humans can learn about (and reverse!) biodiversity losses faster than we currently are. Ann knows many people who have watched birds for only about 20 years, just a blink of evolutionary time, and many have noticed worrisome changes in just those two decades.[113] We're running out of time to act with any hope of success. So, what can we do? Start with Seven Simple Actions to Help Birds at birds.cornell.edu.

> While visiting the banding station, I saw a fox sparrow that had been captured every year for five years! That hooked me. The experience fundamentally changed the way I thought about conservation. I had no idea how important specific sites were to individual birds. Then I learned that the same hummingbird can fly to and from Mexico eight times over eight years and come directly back to my feeder every time. Incredible! I couldn't help but want to know more!
>
> —**ANN NIGHTINGALE**, Rocky Point Bird Observatory

Engaging and Taking Action

At the other end of the size spectrum is a hummingbird. If you have not reviewed that chapter yet, have a look at its Engaging and Taking Action section. You will read about the Rocky Point Bird Observatory (RPBO), an organization that has trained many volunteers and always requires committed assistants. Many of the same actions that apply to small birds also apply to large birds, and at RPBO volunteering can range from helping an

AS CARRION EATERS, VULTURES ARE OFTEN FOUND ON THE GROUND BESIDE AND FEEDING ON DEAD ANIMALS. IT IS THEN THAT YOU CAN GET A SENSE OF THEIR LARGE SIZE.
PHOTO ROB ALEXANDER.

experienced bander record data, banding birds, managing or mending mist nets, or interpreting and encouraging others to have an affinity for birds. RPBO runs both in-person and online events, and could always use help with Zoom presentations, book and retail sales, and workshop organization. That said, they don't actually band Turkey Vultures, as they have several traits that make them difficult to work with (vomiting and defecating on their legs are two of them!). But many other birds are banded, and some can be tracked. Volunteering your time to support the science around birds is rewarding, and it's hard not to make an emotional connection when you have a tiny bird in your hand. Other local opportunities include Christmas bird counts, breeding bird surveys, and crowd-sourced databases that track window strikes (when birds fly into windows) or numbers and locations of backyard birds. A great place to begin your search for something that fits your schedule and the number of hours you have is Birds Canada.

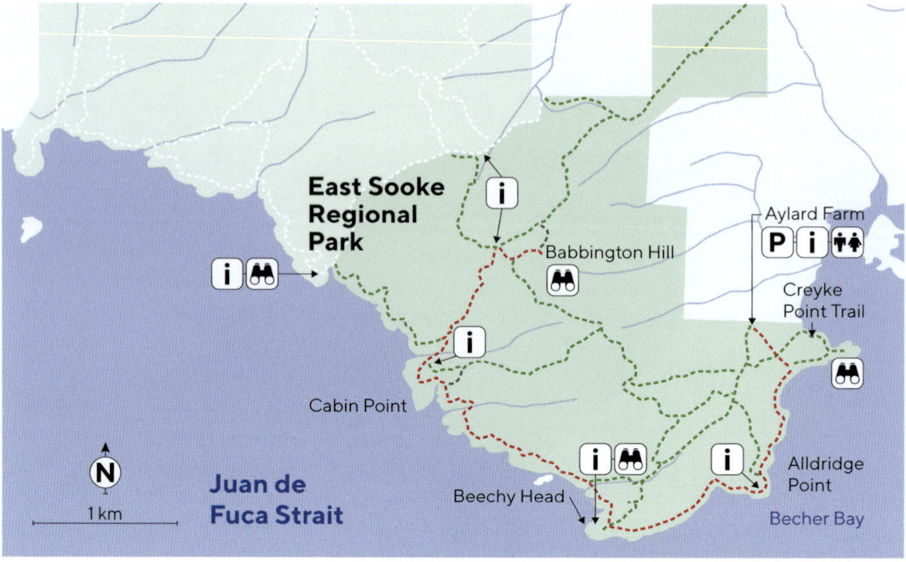

FOLLOW THE EAST SOOKE REGIONAL PARK TRAIL FROM AYLARD FARM PARKING AREA TO BEECHEY HEAD ALONG THE COASTAL TRAIL FOR CLEAR VIEWS OF JUAN DE FUCA STRAIT AND THE VULTURES AND OTHER RAPTORS THAT RIDE THE RISING AIR COLUMNS. SOURCE: CRD EAST SOOKE REGIONAL PARK.

Minimizing Your Carbon Footprint

The Lochside Regional Trail and Galloping Goose bike paths are two cycle routes I highlighted in my first book, *Active Vancouver: A Year-Round Guide to Outdoor Recreation in the City's Natural Environments*. To me these are among the most stunning coastal paths in BC. The Lochside Regional Trail is a 29 km cycling trail between the Swartz Bay ferry terminal and downtown Victoria. The Galloping Goose Regional Trail is a 55 km multi-use trail running between Victoria's Johnson Street Bridge and the Sooke River valley. From downtown Sooke, it's an additional 22 km ride to East Sooke Regional Park. I can't imagine a better autumn cycle excursion on the BC coast, so it's certainly worthy of setting a summer goal to ride that distance. To make this goal available to more potential cyclists, it's good to know you can travel this route in one direction, and then return on the BC Transit #61 bus that runs between Victoria and Sooke. That also means it's possible to transit to Sooke and consider other ways of getting to East Sooke Regional Park from downtown Sooke.

Directions and Accessibility

Take the BC Ferry from Tsawwassen to Swartz Bay, and carry on to Highway 17 (or from Horseshoe Bay to Departure Bay via Highway 1). Locate Highway 14 and take it to Sooke Road, turning right onto Gillespie Road. Turn left on East Sooke Road, then right on Becher Bay Road to reach the Aylard Farm entrance to the park. There is no public transit to this entrance.

There are accessible toilets at each park entrance and at the end of the Iron Mine Bay / Coast Trail, but otherwise the trails are rough and rocky. Contact Sooke – Port Renfrew Tourism for more information.

Other Locations to View Turkey Vultures

1. Turkey Vultures are found throughout Southern BC in open country, especially within a few miles of rocky or wooded areas. With a changing climate, these vultures are expanding their range northward and are becoming more common in locations like Quesnel and Prince George. The best time to see them soaring is on warm summer days when the sun heats the land and the air rises, mostly between 10:00 a.m. and 2:00 p.m. While driving through the Fraser Valley, look for them using the thermals created by steep topography, for example by Sumas Mountain.

2. Other locations to view Turkey Vultures are the high points on the Gulf Islands, such as Mount Maxwell Provincial Park on Salt Spring Island, Mount Parke on Mayne Island, Mount Warburton Pike on Saturna Island, Bluffs Park on Galiano Island, and Oak Bluffs Park on North Pender Island.

3. Turkey Vultures also breed and feed in the area, so it's not uncommon to count ten or 12 vultures in both Sooke and Victoria on the Christmas bird counts, so you could conceivably see them on the South Island throughout the year.

Other Area Interests

1. Sooke Potholes Provincial Park is a great summer cool-off spot.

2. Roche Cove and Matheson Regional parks offer good birding and scenic trails.

3. Galloping Goose bike trail is a carbon-neutral way to get to within 22 km of East Sooke Park.

4. Whiffin Spit Park is a fantastic picnic and wildlife viewing area, an easy walk, and accessible to wheelchairs for a good portion of it.

5. Kayaking Sooke Harbour is an option for additional shoreline viewing of wildlife.

Additional Resources

1. **Cornell Lab** offers a suite of great resources, including Merlin ID Bird, Allaboutbirds.org, and Bird Academy.

2. **The Rocky Point Bird Observatory** on southern Vancouver Island doesn't do banding on Turkey Vultures, but they do on many other species. If you have interest in banding birds, check them out.

3. **Tibetan Sky Burials** – Though it was beyond the scope of this chapter, my research opened a circle of life story that was new to me. As with so much internet research, it is easy to go down "rabbit holes," and this story has to do with a gift to nature performed by Vajrayana Buddhists on the Tibetan plateau. It's the practice of sky burials on mountaintops, releasing a body to the vultures and other scavengers.

CLIMATE CHANGE AND TURKEY VULTURES

Climate change will affect most bird species detrimentally, but for the Turkey Vulture it seems to be a neutral factor, or maybe even a positive, as it seems to be extending their northern range. Turkey Vultures are more resilient than most birds, as they are relatively adaptable in both their diet and their nest sites and, being detritivores (meaning they scavenge on the dead of other species), their needs are not particular to a certain geographical area.

A COLOURFUL MALE CHUM SALMON READY TO FERTILIZE A FEMALE CHUM'S EGGS IN THE SHALLOW WATER OF A STREAM. PHOTO: MICHAEL SCHMIDT PHOTOGRAPHY

Chum, Coho, & Chinook Salmon

in Goldstream Provincial Park, Vancouver Island

AT A GLANCE

WHAT: Chum Salmon (*Oncorhynchus keta*), and possibly Coho (*Oncorhynchus kisutch*) and Chinook Salmon (*Oncorhynchus tshawytscha*)
WHERE: Goldstream Provincial Park
WHEN: Late October to late November
TRADITIONAL TERRITORY: Songhees, Esquimalt, W̱SÁNEĆ, Coast Salish
ACCESS: Easy

Salmon are probably among the species most symbolic of coastal British Columbia. It may be because they're so widespread; almost every major coastal river and stream has held a salmon run. Perhaps it's because most families have a relative who has fished recreationally for salmon. Or maybe it is due to the central role they've played historically: For millennia, First Nations communities have relied on salmon for nourishment, to the extent that salmon have become an integral part of their cultures and spirituality. Early European settlers also learned to rely on salmon as an abundant seasonal food resource. Whatever the reason, coastal dwellers may have a hard time naming five species of local waterbirds, but less problem naming our five species of Salmon: Coho, Sockeye, Chinook, Chum, and Pink.

Goldstream Provincial Park on southern Vancouver Island harbours one of nature's most fascinating spectacles: the return of spawning Chum Salmon to the natal stream in which they began their lives, three or four years before. The Goldstream River spawning run is close to the ocean, but some salmon migrate more than 3000 km (up to 50 km per day!) upward through BC's freshwater rivers. The Goldstream River is nestled within an

WITH ITS CHARACTERISTIC GREY-GREEN BODY BROKEN BY REDDISH-PURPLE VERTICAL BARS, AND LONG HOOKED JAWS, THIS MALE CHUM SALMON SHOWS ITS FULL SPAWNING COLOURS. PHOTO MARION SHIKAZE.

old-growth forest of Douglas Fir and Western Red Cedar, leading out to the estuary and Finlayson Arm at the southern end of Saanich Inlet, and ultimately connecting to the Strait of Georgia. BC Parks has created fully accessible riverside trails and observation platforms from which to view the salmon run, and the best time to visit is from late October to late November. Chum Salmon, a **keystone species** (a species whose role within an ecosystem has a disproportionate effect on other organisms within that system), are the primary draw. Female Chum have a narrow body with a characteristic dark red horizontal bar down the middle of their bodies, while males are larger with grey-green bodies broken by reddish-purple vertical bars, long hooked jaws, and dog-like canine teeth (which is why Chum Salmon are often called Dog Salmon). In some years, there are also considerable numbers of Coho (with dark green heads and backs, and cherry red on their sides), but they tend to be shy and secretive, and often hide beneath logs or overhanging banks. Much larger Chinook Salmon (also known as Spring or King Salmon) are, encouragingly, increasing in numbers on the Goldstream River after years of low turnout, possibly due to habitat restoration activities by volunteers. All of these species spawn in the gravel beds of the river, just as their species have done for millennia, but like other species that have evolved together, they each have their own preferred niche habitat, and therefore spawn on different sections of the river. The Chum Salmon spawn on the lower portion, generally preferring stream sections close to the estuary; the Coho Salmon spawn above the Finlayson Arm Road bridge; and the Chinook Salmon tend to travel up the river as far as they possibly can.

The Goldstream River spectacle is really all about the *cycle of life*. Salmon begin and end their lives in the same gravel bed. The female Chum uses her tail to dig a nest (called a "**redd**") in the river's gravel bed, up to half a metre deep! Before covering over the hole for protection, she lays about 400 eggs, also known as "roe," in the depression, while male Chum (usually several at once) immediately fertilize them with their sperm, known as "milt." There's a great video online of this spawning action titled *Chum Salmon Spawn!* from local Bowen Islander Bob Turner. The female Chum continues the process of digging multiple redds until she has laid all of her approximately 3,000 eggs. Less than 1 per cent of her eggs will survive to produce the next generation! With her life's work complete, she will die shortly after, the nutrients of her body recycled into the wider environment, including the

THE ALEVIN IS AN EARLY STAGE IN A SALMON'S LIFE CYCLE, A TIME WHEN IT STILL HAS A YOLK SAC TO SUPPLY IT WITH NUTRIENTS. PHOTO OPENCAGE (DAIJU AZUMA), WIKIMEDIA COMMONS, CC BY-SA 2.5.

ETHEREAL REFLECTIONS CAN IMPART A WATERCOLOUR EFFECT TO THE SALMON UNDERWATER. PHOTO MICHAEL SCHMIDT PHOTOGRAPHY.

next generation of salmon.[114] The life cycle of the new generation begins in a few months when the fertilized eggs become "**alevins**" (appearing as very small fish with a nutritious yolk sac on their underside that sustains them in the safety of the gravel bed for the winter); then "fry" (salmon babies that emerge above the gravel in search of tiny food, mainly plankton); then "parr" (which are able to eat larger invertebrates, such as aquatic insects). By the following summer, the babies have grown to be finger-sized juveniles known as "smolts" and have moved down the river and closer to the estuary, a transition zone, where they begin to live in "brackish" water, a mix of fresh and salt water. It's in the estuary that Chum smolts imprint on the chemical signatures of their home stream before entering the Strait of Georgia, and ultimately the open ocean of the Pacific, where they will travel thousands of kilometres, growing to approximately 6 kg, before returning to their natal streams to spawn.

When Chum return to their natal streams, many predators will be waiting for them; they are such a reliable seasonal food source, a coastal nutrient backbone, that other creatures' natural cycles have aligned around these salmon, taking advantage of the abundance of food they provide. Consider the diversity of species that prey on the salmon. Black Bears, Grey Wolves, and Killer Whales are all top predators whose movements adjust with the salmon run. American Dippers feed on the eggs not properly covered by gravel, and so do Cutthroat Trout. Insects gather to feast on the carcasses, which are also a Bald Eagle's primary food source. Loons feed on fry, and so do Great Blue Herons. (In fact, one research study that tested the guano of Great Blue Herons showed that herons may have the most significant impact on salmon fry.)[115] The health of this diverse food web depends on the availability of water: clean, cool, and lots of it!

Healthy river systems and watersheds are not a given, especially those in places where humans see a valuable resource to be exploited. Logging has, in the past, destabilized soil structure along riverbanks, causing erosion, and in turn silting up critical salmon habitat. Healthy forests provide shade to keep the water cool and help cycle nutrients into rivers and streams. Estuary, wetland, and watershed degradation greatly influence the survival of the next generation of salmon. The changing climate too, with its longer, drier summers, warming waters, changing patterns of precipitation, and rising ocean temperatures, can be detrimental to salmon. (Consider that warmer water holds less oxygen for salmon and other gilled creatures to breathe.) Try a little investigation. Find a satellite map of Vancouver Island, Metro Vancouver, and the Fraser Valley. Focus on where the rivers are, and notice the overlap of human activities that surround them. Pay attention to the construction of roads near rivers, and consider the corresponding tire rubber that is shed from automobiles driving next to them. See the suburbs built next to the rivers and the tree canopies that don't exist anymore because of the new homes. Look for the logging clear-cuts uphill of the rivers, especially on Vancouver Island. Notice southern Vancouver Island's, or the Fraser Valley's, agriculture and irrigation, and the dikes used to channel the river's natural flow. These *all* impact salmon, and activities like these may be slowly choking the natural dynamics of rivers and the biodiversity that depends on them.

Rivers have above-ground components that we see and experience as we walk beside them on dikes, paddle on them in our boats, or fish or swim near their shores, but their greater complexity lies in the interaction between what we see on the surface and the unseen currents that move below the ground. New modelling suggests that the vast majority of water in a river system moves slowly through a labyrinth of networks under the gravel, cobble, and sand, and that this **subterranean river habitat** is more biologically productive than previously understood. This under-river matrix is home to microbes and aquatic insects critical to the foundation of a river's food chain, and just as it filters organic material within the river, it also releases nitrogen, phosphorous, and other nutrients throughout the system, making these available at the surface. The model likens this complex interaction to an "immune system," and suggests that the myriad of human activities weaken the system, making the aquatic organisms more vulnerable to stresses such as fish-killing parasites. Add climate change and consider how these stresses affect organisms up the food chain, from subterranean aquatic insects, into salmon, and ultimately into predators such as Bald Eagles and Grey Wolves. So as you consider the Chum Salmon of the Goldstream River, reflect on all of the bits and pieces we humans take from the river, and on how our activities alter not just the surface but also the unseen alluvial matrix underneath, and ultimately the river's overall resilience.[116]

Walking the trails of Goldstream Provincial Park in the late fall is a multi-sensorial experience. You can smell the pungent aroma of a few rotting Chum Salmon carcasses (hopefully somewhat masked by the earthy smell of decaying autumn leaves), while listening to the riffling stream interspersed with definitive splashes or leaping salmon as the males posture for position around the egg-laying females. If you watch quietly, move slowly, and don't disturb the fish, you'll likely see Chum Salmon working their way upstream – some in pairs, others in groups, and if you're lucky you may see a female actively digging or defending a redd.

A few "Salmon Come First" watching tips from BC Parks include:[117]

→ Avoid wearing brightly coloured clothing, especially reds, purple, and pinks, which salmon see very well.

→ Bring sunglasses or cameras with polarizing lenses to cut down on glare from the water. These are especially good for children.

→ Start your visit at the Nature House (Freeman King Visitor Centre) to learn more about the salmon or the eagles and other birds of prey attracted to the park's salmon run. There may also be free interpretive programs at the centre.

→ Try to find viewing sites that are up high on the top of banks or on the bridges, to increase visibility, while staying on the trail and away from traffic.

→ Leave your dog at home. If you do bring a dog, it must be on a leash and kept out of the river.

→ Bring your binoculars as there will be plenty of other wildlife around to see, including Bald Eagles.

→ And remember the weekends and afternoons will likely be crowded, so visit outside of those times if you can.

GOLDSTREAM PROVINCIAL PARK TRAILS ARE WIDE, EASY TO WALK, AND MOSTLY IN COOL AREAS, EITHER NEAR WATER OR UNDER TALL TEMPERATE FOREST TREES.

MUCH OF SOUTHWESTERN BC IS CLASSIFIED AS TEMPERATE FOREST. ***TAKE TIME TO FIND A BEAUTIFUL PATCH OF IT, AND THEN PAUSE TO SIT, WATCH, AND LISTEN.***
PHOTO ROB ALEXANDER.

While at the park, wander the trails to find the old-growth (600-plus-year) Douglas Firs or the park's own 47 m high Niagara Falls, plunging into a crystal-clear canyon pool. These streamside riparian areas are also enriched by the Chum, as Black Bears drag salmon carcasses away from the river to eat, moving ocean nutrients from the streamside into the forest.

Another important element to reflect on while viewing salmon is that you are standing on the unceded traditional territory of the Songhees and Esquimalt Nations and the W̱SÁNEĆ (pronounced wh-say-nuch) peoples, represented by the Tsartlip (Sart-Lip), Pauquachin (Paw-Qua-Chin), Tsawout (Say-Out), Tseycum (Sigh-Come), and Malahat (Mal-a-hat) Nations.[118] Indigenous People not only relied (and rely) on salmon for nourishment, they also sought spiritual guidance from them. Historically, BC coastal First Nations carried out salmon ceremonies, during which they welcomed the salmon back to the rivers and tributaries they had been caring for throughout the year, and many continue the practice today. The bones of the first salmon caught would be returned to the ocean as a sign of gratitude and hospitality so other salmon would come and give their lives to the people. Today, we can all take a lesson from coastal BC's First

Nations people, giving our gratitude for the returning salmon, and through our stewardship, make our rivers, streams, and wetlands a hospitable place for them to return.

> Suddenly a smoke-like cloud appears from the nest; that's gotta be male sperm. So the female must have just dropped some eggs. I cautiously move my camera forward to the side of the nest. Across the nest, several males hover.... Then in a remarkable moment she bellies down to the gravel, gapes, and males rush to her side.
>
> —**BOB TURNER**, YouTube videographer, *Chum Salmon Spawn!*

Engaging and Taking Action

Federally run **salmon enhancement facilities** (known ordinarily as salmon "**hatcheries**") are common in many regions. They are set up to capture a small portion of spawning wild salmon, acquire the females' eggs and the males' milt to produce fertilized eggs, and then care for the salmon young through to their release at the smolt stage. This increases the survival rates of juvenile salmon significantly. As you might expect, human management of fish has its supporters and detractors. On the one hand, hatcheries release salmon into a food web that includes seals, sea lions, and Orcas, as well as recreational fishers. It is estimated the Orcas alone eat 25 kg of salmon per day, meaning BC's approximately 400 Northern and Southern Resident Killer Whales probably consume 1000 tonnes of salmon each year![119] The expectation is that ultimately more salmon will return to the river, offsetting past losses due to environmental stressors. Detractors of this type of fish management suggest that hatchery fish compete with wild fish for increasingly scarce resources and that hatcheries domesticate salmon by selecting for genes suited to the hatchery environment but maladaptive in the wild. This results in reduced fitness of hatchery salmon once released.[120] Whatever side you fall on, it's indisputable that human activities have significantly impacted the millions of salmon that forge their way up the myriad streams to spawn on the west coast of British Columbia. Fortunately, there are many ways to help improve the situation for salmon. Most hatcheries have opportunities for volunteering, through education programs, stream monitoring, or the capture and release of fish, for example. Opportunities for habitat restoration also exist through some hatcheries, local Streamkeepers organizations, the Watershed Watch Salmon Society, the Pacific Salmon Foundation, or local non-profits such as the Squamish River Watershed Society.

A WAY TO MINIMIZE YOUR CARBON FOOTPRINT IN GETTING TO GOLDSTREAM PARK IS TO CYCLE THE LOCHSIDE AND GALLOPING GOOSE TRAILS. PHOTO BRAKELIGHTFOTO.

The Fraser Valley Conservancy (FVC) is a charitable organization that protects land and water for future generations. Focusing on land protection, stewardship, and habitat enhancement, it seeks to promote the acquisition and conservation of ecological and historic landscapes in the Fraser Valley. It offers many opportunities to help residents engage in local projects and to take action in their community. The FVC takes an active role in stewarding salmon habitat. The estuarine habitat also supports the endangered Western Painted Turtle. If you are interested in this species, see Other Area Interests at the end of this chapter.

Minimizing Your Carbon Footprint

Salmon viewing is a family-friendly activity, so load up a vehicle with many passengers, especially youth, and make your way to Goldstream Park as a group.

If the late autumn weather is favourable, consider a cycle on the Lochside and Galloping Goose regional trails with an overnight at Goldstream Park campground.

Directions and Accessibility

Goldstream Provincial Park is located 16 km northwest of Victoria on southern Vancouver Island on the Trans-Canada Highway (Highway 1). There are separate entrances for the main campground and the day-use area. The entrance for the day-use area is at the junction of Highway 1 and Finlayson Arm Road.

Goldstream Provincial Park has a page for the areas that are fully accessible.

Other Locations to View Salmon

1. Salmon are likely swimming in your neighbourhood! If you live in Metro Vancouver, locate your local streams and organizations that are dedicated to improving salmon habitat. Unfortunately, the city of Vancouver has lost most of its salmon streams due to urban sprawl – out of about 50 streams, only two remain. I live in the Tri-Cities, and there are several great locations to watch salmon, including Noons Creek, Hoy Creek, Mossom Creek, and Coquitlam River. The City of Coquitlam even publishes a family-friendly "Salmon Come Home" map to guide residents of the community.

2. The Watershed Watch Salmon Society provides a list of spawning sites and times, which includes southern Vancouver Island, Metro Vancouver, and the BC interior.

3. Perhaps the most significant tourist attractions for spawning salmon are the Capilano River Hatchery in North Vancouver and the Weaver Creek Spawning Channel in Harrison Mills, but my favourite Greater Vancouver location is the Seymour River Fish Hatchery & Education Centre on the Seymour River, as it is surrounded by beautiful old-growth forest.

METRO VANCOUVER AND SOUTHERN BC HAVE MANY LOCATIONS TO VIEW WILD SALMON. AMONG THE MORE PHOTOGENIC SALMON TO SEEK OUT ARE SOCKEYE WITH THEIR BRIGHT RED SPAWNING COLOURATION. PHOTO MICHAEL SCHMIDT PHOTOGRAPHY.

Other Area Interests

1. An additional side trip to another ecosystem consisting of Garry Oak and Arbutus is the hike to the summit of Mount Finlayson, for views of the straits of Georgia and Juan de Fuca.

2. Western Painted Turtles, a BC red-listed species, can be found 17 km away at Swan Lake–Christmas Hill Nature Sanctuary. This is BC's only *native* turtle (others such as Red-eared Sliders or Mississippi Map Turtles are introduced species, released from former pet owners). To view the Western Painted Turtles, begin on the boardwalk that crosses the lake and watch for them sunning themselves on the logs by the shoreline. They do this in early spring (mid-March to early May), after coming out of hibernation. By late spring too much vegetation conceals them, and the air gets too warm, so they spend more time in the water and aren't so easy to spot. Habitat enhancement is important, and saving wetlands is critical for these turtles; so is controlling Bullfrogs, which feast on the juveniles. The presence of automobiles around the periphery of turtle habitat also leads to roadkill. These mostly vegetarian creatures (interestingly, they are more carnivorous the farther north you find them) eat aquatic plants, and can live up to 50 years! Because turtles are generally long-lived, the turtle researchers I spoke with all suggested not to consider any kind of turtle as a pet. They all grow large, are susceptible to diseases, and become increasingly difficult to care for properly.

   Western Painted Turtles, being an endangered species, provide a good reason to get involved in wetland restoration. If you do come across one, never move them from one habitat to another; only move them when they are on roadways (and always in the direction they were moving in the first place). They often end up crossing roadways in June, their breeding season. Because they are rare, consider being a citizen scientist by taking pictures of their shell, their head, and a side view, then post your images to iNaturalist. And continue to learn more about them, starting with a great website on the Reptiles of British Columbia or the BC Environment brochure titled *Wildlife At Risk*. For hands-on experiences, consider the surveys and stewardship activities led by the British Columbia Marsh Monitoring Program.

3. The Butchart Gardens is not too far from Goldstream Park and, if you've read the chapter on insects, you'll know there are opportunities to see late season dragonflies, bees, or butterflies.

WESTERN PAINTED TURTLES ARE A RED-LISTED SPECIES, MEANING THEY ARE ENDANGERED. ALTHOUGH SEVERAL LOCATIONS HAVE THEM, INCLUDING VANCOUVER ISLAND'S SWAN LAKE NATURE SANCTUARY, WILDLIFE OBSERVERS ARE ENCOURAGED TO NOT DISTURB THEM AND NEVER TO MOVE THEM. PHOTO ROB ALEXANDER.

4. Although Turkey Vultures have mostly departed southern Vancouver Island by late October, you are not too far from the beautiful natural landscape of East Sooke Regional Park.

5. There are a number of companies that rent out kayaks in and around Victoria Harbour for those interested in doing some wildlife observing from the aquatic realm.

Additional Resources

1. Search the internet for the **Pacific Salmon Explorer**, a data-rich visualization site that allows you to zoom into a specific river system, see the salmon that inhabit it, along with their habitat and population status, and understand more thoroughly the pressures that exist for those specific salmon in that particular region.

2. Follow one of the Raincoast Conservation Foundation's flagship projects, "**Habitat Restoration in the Fraser River Estuary**." Their five-year restoration project is designed to create openings in several of the man-made barriers that restrict the passage of juvenile salmon on their natural migration.

3. Check out the video ***Pacific Salmon and Their Circle of Life*** at the oursalishsea.ca site hosted by Bob Turner.

CLIMATE CHANGE AND PACIFIC SALMON

The changing climate is leading to longer, drier summers, late and potentially heavier autumn rains, and rising river and ocean temperatures, all of which are detrimental to salmon. Warmer water holds less oxygen for salmon and other gilled creatures to breathe. Prolonged summer droughts lead to low river flows and too-warm water during spawning. And when prey species, like the aquatic insects in rivers and the plankton and small fish in the open ocean, are also stressed by climate change, the effects are felt by species up the food chain, including salmon.

A BEAUTIFUL PORTRAIT OF A MALE CALIFORNIA BIGHORN SHEEP, PHOTOGRAPHED IN CHASE, BC, BY ERIC GROSS.

California Bighorn Sheep

at Lac du Bois Grasslands Protected Area, Kamloops

AT A GLANCE

WHAT: Bighorn Sheep (*Ovis canadensis*)
WHERE: Lac du Bois Grasslands, Kamloops
WHEN: Mid-October to mid-November
TRADITIONAL TERRITORY: Secwepemcúl'ecw (Secwépemc), Nłeʔkepmx Tmíxʷ (Nlaka'pamux)
ACCESS: Moderate

Even a casual viewer of nature documentaries has likely seen the magnificent display of strength and agility shown by male Bighorn Sheep (rams) vying for dominance and access to females (ewes) during the autumn rut. This is a time of prime physical condition for these males, which stand about a metre high at their shoulders. Dominant males pair up, back off one or two metres, then charge, jump, and lunge at each other, smacking horns at full force, with horns that can weigh up to 14 kg (30 lb.). The horns themselves are equivalent in weight to all the bones in their body![121] The curvature of the horn bone structure takes most of the shock energy, protecting them from brain damage! This unforgettable sight happens in many places in BC, and it's from this province that many of these documentary films get their footage. One of the great places to see the iconic California Bighorn Sheep is on the Battle Bluff hike in the Lac du Bois Grasslands Protected Area, adjacent to Kamloops Lake.

Typically, the bighorns are light to chocolate brown in colour, and both the ewes and rams have white rumps. The autumn rut determines which males get to breed with the ewes. Add a six-month gestation period for the ewes after breeding, and that leads to new lambs, usually one lamb per ewe, born in May or June. These hoofed mammals and their young are

THE HORNS OF CALIFORNIA BIGHORN SHEEP SHOW PROMINENT RIDGELINES, EACH KNOWN AS AN ANNULUS, DEFINED EACH WINTER WHEN GROWTH SLOWS. THESE ARE CLEARLY SEEN ON THIS MALE. PHOTO ERIC GROSS.

well adapted to country with rugged, steep, rocky cliffs, and are often found on south-facing drier slopes with lower snowpack in the winter, allowing them to graze all year. The rocky, bluff-like locations allow them to seek cover from predators, which is especially important after their lambs have arrived. A key adaptation to cliff environments are their hooves, which have a hard outer edge surrounding a softer concave middle, giving excellent traction on rocky terrain. Their horns grow throughout their life and produce a prominent ridgeline (known as an annulus) each winter when growth slows. These lines, which are analogous to tree rings, give biologists a convenient means of assessing the ages of individuals in the herds. Bighorns typically live for ten to 12 years, but as a species, they've been around for over 600,000 years – a long time indeed![122] See Appendix C regarding Evolutionary History of Selected Species Covered in *Wildlife Weekends*.

Bighorn Sheep have practised their autumn rut ritual for eons, as they are known as a species that originally entered North America from Siberia, crossing over the **Bering land bridge**, in the **Pleistocene Period**. At one time, millions likely roamed throughout most of North America, ranging as far south as central Mexico. Glaciation separated bighorns into two isolated ice-free areas (known as "**refugia**"). One area was in central Alaska and the other south of the Columbia River, in present-day Oregon. The Alaskan sheep evolved into Thinhorn Sheep (also known as Dall and Stone Sheep), seen today in northern BC, Yukon, and Alaska. Sheep south of the Columbia River grew more heavy-horned, apparent in today's Rocky Mountain, California, or Desert Bighorn Sheep. By 1800, the total continental population was estimated to be no more than 200,000, while the California population of bighorns today are a fraction of that, at just 10,500.[123] There are myriad historical reasons for their population decline, and these include past unregulated hunting, loss of habitat due to urban and agricultural development, overgrazing of rangelands, introduction of lower-nutritional-value invasive species, vehicle collisions on highways (watch for Bighorn Sheep crossing signs when driving), all-terrain vehicle damage to remaining habitat (especially harmful during lambing season), and diseases contracted from domestic sheep (domestic sheep are related to wild sheep by a common ancestor from Asia). The principal bighorn decline happened between 1850 and 1900.[124]

SUMMER VIEWING OF BIGHORNS OFTEN INCLUDES SEEING THE LAMBS, BORN BETWEEN LATE MAY AND MID-JUNE. IT'S ALSO A TIME WHEN THE SHEEP ARE ESPECIALLY VULNERABLE, SO VIEWING AT A DISTANCE IS ESPECIALLY IMPORTANT. PHOTO TERRY CARR.

Today, hunting in BC is heavily regulated and therefore doesn't impact BC's population significantly, but one thing that still does harm wild sheep is contracting a bacterial disease from their domestic cousins. Edyta Marcisz is a wildlife researcher and master's student at Thompson Rivers University, and I had the opportunity to speak with her about what she calls "the biggest issue affecting wild sheep in BC, today." **Movi** (*Mycoplasma ovipneumoniae*) is a pneumonia-causing bacterium commonly carried by domestic sheep, but the domestics have developed a resistance to it. The pathogen transfers easily between domestic sheep and bighorns (all it takes is nose-to-nose contact when foraging on grasslands or at a water source), consequently causing a respiratory disease in bighorns that is usually lethal. Wild sheep often die within days, but as they are highly social animals, those days after infection also mean further interactions within the remaining wild herd, and the potential for a widespread mass die-off is significant. That happened in 2013 when the Chasm Bighorn Sheep herd was exposed to Movi, and declined 75 per cent, from about 100 to 25! The pathogen gets into the lungs of Bighorn Sheep and leads to pneumonia. It is often the ewes that pass it on to their lambs, which in turn don't survive. Lamb survival (also called "**recruitment**") is, of course, key to keeping a healthy population.[125]

If this all sounds similar to COVID-19, it is, as Movi is also transmitted by airborne particles, by coughing or sneezing, and is tested for by nasal swab. But what's different for the sheep is that there is still no effective vaccine available. The best solution is – you guessed it – "social distancing." Wild and domestic sheep populations must be completely separated to avoid disease transmission, and there's a BC Sheep Separation Program to advocate for this. The aim is to educate farmers of domestic sheep to manage with Movi in mind. Sixty per cent of Canada's Bighorn Sheep populations are found in the Fraser River region between Lillooet and Junction Sheep Range Provincial Park near Williams Lake, and the region overlaps with many domestic sheep farms. Unfortunately, excluding domestic sheep from private land is almost

THE GRASSLANDS CAN BE IN STUNNING BLOOM IN MID-JULY IF THE RAINS ARE CONSISTENT. IT'S A BEAUTIFUL TIME TO VISIT, *BUT BE ON THE WATCH TO AVOID WESTERN RATTLESNAKES WHEN WALKING THROUGH GRASSLAND AREAS.*

impossible, as the BC Farm Practices Protection Act (also called the Right to Farm Act) is a powerful piece of legislation that guarantees opportunities to house livestock (including domestic sheep) on private land. Currently, high-risk areas are being identified, and management techniques such as refuge pastures and perimeter fencing are being implemented. I think the most unique idea the Sheep Separation Program has put forward is to create a market for "bighorn-friendly" lamb (similar to dolphin-friendly tuna)! Fortunately for the Lac du Bois Protected Area Kamloops Lake herd, strong management systems do exist.

The Lac du Bois landscape is composed of several protected areas, including the Tranquille **Ecological Reserve**, Rattlesnake Bluffs Conservation Area, and the Dewdrop-Rosseau Creek Wildlife Management Area. These expanses protect an increasingly rare, and at risk, **grassland ecosystem**. Grasslands make up less than 1 per cent of BC's natural environment, but it's an important ecosystem that many animals rely on, and it shelters over 30 per cent of BC's threatened species![126] Rattlesnake Bluffs, a Nature Conservancy site, may look like barren rock, but a closer look at the right time of year (summer to early autumn) reveals signs of life such as a Cliff Swallow colony with their gourd-shaped mud nests, White-throated Swifts in flight above, Yellow-bellied Marmots sunning themselves on the rocks, and Western Rattlesnakes (which typically hibernate by mid-October) moving among the sagebrush. The Tranquille Ecological Reserve protects a representative example of a **Ponderosa Pine Biogeoclimatic Zone**, which archeological evidence shows burned naturally, on average, once every 18 years. Historically, First Nations burned this landscape as well, but 20th-century practices of fire suppression have been so effective they have lengthened the intervals between fires considerably. The unintended result has been that when fires do happen, they happen big! You can see signs of a recent fire on the five-kilometre (return) hike to Battle Bluff.

Battle Bluff is a good area for seeking out bighorns. (Remember to bring binoculars to watch them from a good distance – they're skittish; and pay attention underfoot for the Prickly Pear Cactus, which is well-named!). These bighorns are descendants of sheep that were transplanted here in the mid-1960s. In autumn there are two other species I would encourage you to search for, both of which stop in this area on their migrations south. The first is a flock of American White Pelicans that breed in the BC interior and then feed on salmon near the outlet of the Tranquille River where it meets Kamloops Lake, before migrating south. You can often see these goose-sized birds from the road in October, but it still requires binoculars as they will likely be a fair distance away. The second migrant bird of note is a flock of Sandhill Cranes, often seen near the ponds that abut Tranquille Creek

LOOK FOR GOURD-SHAPED MUD NESTS BELONGING TO CLIFF SWALLOWS AT NEARBY RATTLESNAKE BLUFFS, A NATURE CONSERVANCY SITE. PHOTO TERRY CARR.

CLIFF SWALLOWS HAVE METALLIC, DARK-BLUE BACKS, CINNAMON-COLOURED RUMPS, AND WHITE FOREHEADS. THESE INSECT EATERS ARE LONG-DISTANCE MIGRANTS, VENTURING TO MEXICO, CENTRAL AMERICA, AND AS FAR AS EASTERN SOUTH AMERICA IN OUR WINTER. PHOTO ROB ALEXANDER.

IF VISITING THE LAC DU BOIS GRASSLANDS IN THE FALL, *WATCH ALSO FOR SANDHILL CRANES RESTING WHILE ON MIGRATION.* PHOTO TERRY CARR.

Road in October to early November. For both species, find a safe high-point to pull off, near where Tranquille Creek Road meets Tranquille-*Criss Creek* Road, which then passes over the Tranquille River. This entire area is an at-risk BC ecosystem, so please stick to hiking trails and roadways, as veering away from these leads to erosion of the ecosystem's fragile soil, and has the potential of spreading introduced species (from plant seeds stuck between the treads of tires or boots) into this special area. In addition, Bighorn Sheep are affected by our presence, and even more so if we bring dogs with us, so please leave your dog at home.

This region was devoid of bighorns for most of the 20th century, until they were reintroduced in the grasslands north of Kamloops Lake in the mid-1960s (and again in the 1980s). I had an opportunity to walk this area with wildlife enthusiast Frank Ritcey, whose father Ralph was involved in the mid-1960s reintroduction of a herd of 12 sheep, which now numbers about 300. The Bighorn Sheep that range near Battle Bluff are part of the wider Kamloops Lake herd (California ecotype) and are non-migratory. Other sheep herds, however, often migrate between hills (summer) and valleys (winter). Those that do migrate learn to do so culturally, that is, the learning is passed down from parents to their young.[127] This is important to know, because if a sheep is translocated to another area for management reasons, it must be done with care for the young to know where to go!

The Kamloops Lake herd spreads alongside the north shore of Kamloops Lake and up Deadman Creek valley. Other herds exist in the region and southern BC (see Other Locations to View Bighorn Sheep), and some of these can interact if a ram ventures far enough in mating season. Other herds are geographically separated, such as those on the other side of the Fraser River, and while this separation means no disease transmission, it also means no genetic flow. Wildlife managers collect herd health data, and monitor their movements to assess connectivity between herds and their genetic diversity. Frank pointed out that because these animals cooperate as a herd, one of them will always be watching you. They have extremely acute eyesight and can detect the slightest movements over a kilometre away. And somehow they must also be communicating to the others nearby about what they perceive *your* threat level to be. Frank mentioned their eyes' irises are horizontal, giving them panoramic vision – all the better to track every movement of a wildlife observer, so if the sheep you want to watch are close to the road, *the best "blind" is your car.*

WHITE PELICANS USE KAMLOOPS LAKE TO FEED ON SALMON AT THE MOUTH OF THE TRANQUILLE RIVER BEFORE MIGRATING SOUTH FOR THE WINTER. PHOTO TERRY CARR.

Rams typically bed down in early afternoon to rest and ruminate. (**Ruminants** are animals with four-chambered stomachs, and resting helps in the process of regurgitating their food, chewing it again, and re-swallowing it.) Bighorns are predominantly grazers, with grasses and other low-growing plants such as Wheatgrass, bluegrasses, sedges, the roots of cactus, and other soft-stemmed plants being their primary food. A smaller portion of their diet comes from shrubs such as Saskatoon Berry, Rocky Mountain Juniper, and many species of willow. In spring and summer, they will also seek out mineral licks containing salts, to balance out their diet with minerals they aren't able to get in the plants they eat. Besides Movi, another serious threat to wild sheep is the lack of good quality winter food. When they share range with domestic sheep, cattle, horses, or other wild grazers such as elk, there may be too few winter nutrients for them, leading to malnutrition, which ultimately makes them more susceptible to disease and predation. Frank mentioned an interesting connection between cattle grazed here, Bighorn Sheep, and an introduced game bird called a Chukar. In the past, when the grasslands were overgrazed, weed species flourished and the Chukar, as seed eaters, did well. When the range was better managed, the grasses returned, and they, in turn, drew in the Bighorn Sheep to feed. The grasses displace the weed species and therefore the Chukar don't do as well because there is less feed for them.

Both the autumn rut and migration, plus the new life of spring with its array of grassland wildflowers, offer great opportunities to photograph animals in their home environments; but I can only recommend it if the photography is to be done ethically. I have mentioned the Raincoast Conservation Foundation several times in this book, primarily because I believe in their conservation ethic, rooted in science. The society has created a set of Guidelines for Ethical Photography. Though all species in this book are commonly and beautifully photographed, Bighorn Sheep rams with large horns are a particular attraction for photographers, and Frank told me of the many times he has witnessed photographers *chasing down* the animals to get a better picture. Raincoast's guidelines[128] centre on being well equipped with a substantial telephoto lens; trying never to interfere with or alter the behaviour of the wildlife you seek to photograph; being conscious of the amount of time you are in the presence of wildlife; not using flash photography; preventing damage to sensitive ecosystems; and *always* putting the wildlife first. Frank shared some of his experiences as he related how often he sees photographers following sheep, and when he talks to them, they always seem to think *they* are not the ones creating an impact, but rather that others are. The key idea is to be aware of it; we all have an impact, and we always need to ask ourselves how we can minimize that impact while still

YOU NEVER KNOW WHAT WILL HAPPEN WHEN YOU'RE PHOTOGRAPHING WILDLIFE!

PHOTO CHRISTINE GAIO.

experiencing nature. An additional consideration for photography is what you do with the photograph afterward. Posting a special wildlife location online can lead to disastrous effects long after you have been there. For an example of how this "Instagram effect" overwhelmed a site in the Kamloops region, search for the story about the Johnson Lake image innocently posted to social media. A social media picture can become a trophy of sorts for the photographer, but with far-reaching effects.

North American wild sheep have become known around the world as some of the most coveted of big game "trophies." The journal *Conservation Biology* published research led by the Raincoast Conservation Foundation's science director, Dr. Chris Darimont, titled "Large Carnivore Hunting and the Social Licence to Hunt." The study doesn't focus on Bighorn Sheep, but rather on Grey Wolves, Grizzly Bears, and Cougars as objects of **trophy hunting**, and it suggests that surveys and opinion polls in BC show strong opposition to trophy hunting generally. According to the report, the public supports hunting for acquiring food but does not deem killing for trophies to be justified. The public sees *trophy* hunting not as satisfying a hunger for feeding a family but rather as satisfying a hunger for *status*.[129] According to the research paper, public policy is ultimately shaped by societal values, and the questionable ethics of trophy-seeking hunting seem to be leading to the withdrawal of the social licence to hunt, which may ultimately be to the detriment of subsistence hunting as well.

> Bighorn Sheep have a long history of social and cultural importance to the people of British Columbia, we even put them on the Province's official Coat of Arms! They are cautious but very curious, if they feel threatened, they will run away, but then they will turn around and look back at you presenting themselves for a perfect picture opportunity, so have your camera ready!
>
> —**EDYTA MARCISZ**, MSc Environmental Science

Engaging and Taking Action

There are several ways you can engage to help Bighorn Sheep. First and foremost, spread the message to others about the potential for disease transfer between wild and domestic, and if you know a farmer, talk with them and let them know the provincial government pays to test domestic sheep for Movi (it should be mandatory, but it's not). If you are a letter writer, then engage your elected officials to get involved. And if you are in sheep

A BRITISH COLUMBIA MOBILITY OPPORTUNITIES SOCIETY (BCMOS) TRAILRIDER ALLOWS PEOPLE OF ALL ABILITIES TO TAKE PART IN SHEEP VIEWING. *CONSIDER THE WILDLIFE VIEWING LOCATIONS YOU KNOW: ARE THEY ALL ACCESSIBLE?* PHOTO BCMOS.

country and see wild sheep interacting with domestic sheep or goats, call the BC Provincial RAPP line.

For those who want to be more hands-on, there are a number of ground surveys organized by the BC Wild Sheep Society in places like Spences Bridge, Savona, Merritt, or Lillooet.

If photography is an interest of yours, consider entering the BC Nature Photo Contest as an opportunity to post your nature photos.

Minimizing Your Carbon Footprint

With Kamloops being a major city centre, there are bus lines that go between Metro Vancouver and Kamloops. Consider taking a bus to the city and then doing a daily car rental from there to explore the Lac du Bois area. The major roads through the grasslands in the area are also very bikeable for those willing to move through the region at a slower pace.

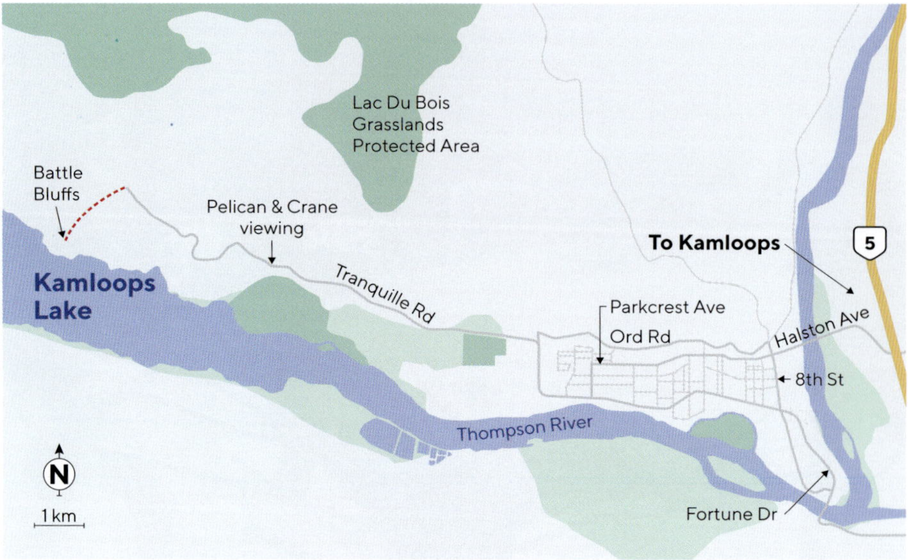

SEE THE DIRECTIONS TO LAC DU BOIS AND BATTLE BLUFFS FOR LOCATIONS TO VIEW SHEEP, PELICANS, OR CRANES IN THE AUTUMN.

Directions and Accessibility

Take Highway 1 to Hope, Highway 5 past Merritt to Kamloops. Once there, cross the Thompson River on Highway 5 and turn left onto Halston Avenue. Turn right onto 8th Street, and an immediate left onto Ord Road, which will turn into Tranquille Road, which in turn will become Tranquille-Criss Creek Road, and then finally it will turn in to Frederick Road, taking you to the start of the Battle Bluff Trailhead.

No trails in this area are fully accessible, but viewing wildlife from the roads is often possible.

Other Locations to View Bighorn Sheep

1. Bighorn Sheep can often be observed while driving between Vancouver and Lytton on Highway 1.

2. Highway 1 between Kamloops and Cache Creek also offers sheep viewing, especially in the alfalfa fields set up for cattle, which often also attract Bighorn Sheep.

3. The Sun Rivers Golf Course just off East Shuswap Road in Kamloops and along the Thompson River as far as Harper Ranch Road are both good locations to view sheep.

4. In the South Okanagan you can also watch for Bighorn Sheep at Vaseux Lake, part of the Nature Trust. Check out the chapter on the Meadowlark Nature Festival, as Vaseux Lake is the site of many of the activities that happen around that festival.

5. Wild sheep in British Columbia are divided into two separate species, known as Thinhorn (Dall and Stone) Sheep and Bighorn Sheep. Rocky Mountain and California are two BC ecotypes of Bighorn Sheep one can see around Kamloops, but to see Thinhorns, you need to drive to northern BC, Yukon, or Alaska.

Other Area Interests

1. Other trails in the area, good for potentially viewing Bighorn Sheep and for experiencing the grassland and Sagebrush ecosystem, are the 3.5 km Mara Loop trail, the 4.25 km Cinnamon Ridge semi-loop trail, and the 5.7 km Dewdrop trail. All offer great views over Kamloops Lake, have moderate hills to climb, and have the plants and animals representative of a grassland ecosystem.

2. In August, consider searching for Common Poorwills in Kamloops or, in May, any of the breeding waterfowl species on the lakes within Lac du Bois.

3. As mentioned, American White Pelicans and Sandhill Cranes are also possible viewing opportunities in the fall as they stop in this area to feed while on their southern migration.

Additional Resources

1. ***Bighorn Headbutting Battle in Canada's Rockies*** – This YouTube video helps us to understand more thoroughly the male headbutting that happens during "the rut."

2. **BC Sheep Separation Program** – Learn more about Movi at this Habitat Conservation Trust Fund site.

3. **BC Wildlife Federation** – This is a "go to" site for so many aspects of BC wildlife, not just Bighorn Sheep.

THE GRASSLANDS AROUND KAMLOOPS LAKE ARE HOME TO CALIFORNIA BIGHORN SHEEP. THERE ARE SEVERAL VIEWING LOCATIONS IN THE AREA TO SEE THESE PHOTOGENIC MAMMALS, INCLUDING FROM THE RAIL ROUTES. PHOTO GILBERTE DIMICOLIS, FLICKR, CC BY-NC-SA 2.0.

4. Raincoast Conservation Foundation – Learn more about **Raincoast's photography ethics** on their website.

5. *Hakai Magazine* article – Though *Hakai Magazine* centres around coastal areas, Annie Roth's article **"When Photographer's Get Too Close, Wildlife Pays the Price"** on photography and ethics is a good read for understanding the issues better.

6. **Hinterland Who's Who** – *Hinterland Who's Who* nature moments was a TV program I grew up with many years ago. During my research I discovered it still exists online! There is a section of the site on Mountain Sheep.

CLIMATE CHANGE AND WILD SHEEP

North America's wild sheep will be affected by climate change, and those in more extreme environments will likely experience the impacts first. Research has found that 30 out of 80 herds of California's Desert Bighorns have already disappeared, and temperature increases and precipitation decreases affecting food sources will likely continue to reduce populations in lower-elevation territory. Thinhorns in northern BC and Yukon will be managing greater increases in temperature, while southern BC's California and Rocky Mountain Bighorns will continue to face the known pressures of a changing climate in addition to habitat loss and fragmentation, domestic sheep diseases, and trophy hunting.

WINTER

ONE OF THE SPECIAL ASPECTS OF WILDLIFE VIEWING IS THE SOUNDS THAT ACCOMPANY THE CREATURES. CAN'T YOU JUST HEAR THIS CROW CAW? PHOTO MARION SHIKAZE

INTRODUCTION TO WINTER

The Fraser River Delta

BY ANNE MURRAY

Author of *A Nature Guide to Boundary Bay* (2006) and *Tracing Our Past: A Heritage Guide to Boundary Bay* (2008), Anne is a lifelong birdwatcher and naturalist who has lived and travelled in many parts of the world. She is the recipient of numerous awards for her work with naturalist and conservation organizations, both nationally and provincially. Find Anne at natureguidesbc.com.

Many years ago, our family moved west from Calgary, driving Highway 1 through the spectacular mountain scenery of the Rockies, Selkirks, and Monashees, and down the Fraser Canyon to Hope. It was mid-winter, bitterly cold and snowing, and the Fraser Valley seemed endless. Snow turned to rain, as the scenery flattened into kilometres of sodden fields, train tracks, and subdivisions. Yet as we drove across the delta farmlands, a pale winter sun sent shafts of light onto Tsawwassen, our new home. It felt promising. On that long-ago day, it would have surprised me to know how very deeply I would come to appreciate this southwestern corner of Canada, and the richness of the flat, rain-washed lands of the Fraser River delta. There is so much wildlife here, even in winter: from great gatherings of Bald Eagles, to vast shorebird flocks creating smoke-like clouds in the sky. Waterfowl congregate on land, sea, and river, Coyotes lope in the marshes, and raccoons wander the suburbs. The estuary's mild maritime climate supports life throughout the winter months.

Observing the seasonal changes through the natural world around us is a deeply satisfying experience. Nature is a combination of many life forms, all in constant flux. Multiple ecosystems, terrestrial and aquatic, meet and merge and mix in the Fraser delta, and it is the coming together of species that provides its astonishing richness. Just as one feels some familiarity with the environment, the unexpected happens: a sudden influx of Snowy

Owls from the Arctic, or a widespread "irruption" of Western Hemlock Looper Moths. Perhaps you spot a tiny Saw-whet Owl sleeping in the drooping branches of a Western Red Cedar, or a small grey stick on the dike suddenly flies off, clicking its wings, to reveal itself as a Carolina Grasshopper.

The words "estuary" and "delta" are often used interchangeably, but the estuary refers to the "mouth" of the river, the stretch of water where salt water and freshwater meet and mingle. The river is tidal as far as 60 km upstream, so the estuary can be considered as that whole stretch of river, together with the sloughs, side channels, and the tidal areas beyond the river's multiple mouths. The estuary receives all the silt and debris accumulated as the river drains three-quarters of British Columbia on its 1378 km journey from the headwaters near Jasper. Over thousands of years, the mouth of the river has been moved steadily west, as the land accreted into the Fraser delta. The river divides into several arms where it meets the ocean, the channels cutting through marshes and wetlands. The channels once wandered freely but were diked and stabilized from the mid-19th century, as was the whole of the delta shoreline. In June, there is a strong flow of melting snow carried downstream by the river, known as "freshet." This remnant of winter snows from other parts of the province is a cloudy plume of silt-laden water that flushes out across the Salish Sea as far as Galiano Island. Boundary Bay, a 16 km wide bay on the south side of the delta, is shielded from much of the Fraser's freshwater outflow by the headland of Tsawwassen–Point Roberts. Some fresh water enters the bay from smaller rivers and streams on the eastern, Surrey side, and the area near Blackie Spit provides attractive winter habitat for waterfowl, shorebirds, and occasional winter visitors, such as Snow Buntings.

The accumulated sand and silt of the present delta, deposited over 10,000 years, produced richly fertile land that is now a mix of farmland, housing communities, industrial areas, and highways, intermingled with cattail marshes, cottonwood stands, coniferous forest, and parks, with the large wetland of Burns Bog in the centre. Together with the extensive tidal mud and sand flats of the estuary, this whole area is a world-renowned habitat for migratory and wintering birds, and is classified as such under numerous international designations. In winter, it is the top birdwatching location in Canada!

The Fraser River estuary is renowned as a migration stopover for up to 50 shorebird species, yet only a few shorebirds stay for the winter. Of those, Dunlin occur in spectacular numbers. These small brown and white birds, with long, probing bills, form flocks in the tens of thousands. They can be seen in winter, feeding with smaller numbers of Black-bellied Plover on the Boundary Bay and Roberts Bank mud flats. As the tide rises, the birds are forced closer to shore. They try to perch on a floating log or uncovered patch of sand, but the high winter tides come right up to the dike. Eventually the birds are squeezed out of available beach space, at which point they take to the air in enormous flocks, twisting and turning until they appear like a distant cloud of smoke draped across the bay. These synchronized flocks, or "**murmurations**," one of the most spectacular sights in the natural world, are a way in which birds evade the pursuit of predatory falcons, such as Peregrine and Merlin. The Dunlin spend hours flying back and forth over the water until darkness falls and the tide drops, exposing wet sand and the possibility of food.

Dabbling ducks and diving ducks, mergansers, grebes, and loons nest in Saskatchewan, Alberta, and the valleys of northern and eastern British Columbia. Dabbling ducks, such as Mallard, American Wigeon, Northern Pintail, and the little Green-winged Teal, are those that upend themselves (heads underwater and bottoms to the sky!) to feed in shallow water. In contrast, the diving ducks, such as Surf Scoter, White-winged Scoter, Bufflehead, and Greater Scaup, prefer deeper water and submerge completely to feed on shellfish. Mergansers are fish-eating specialists, with saw-shaped bills. Grebes and loons raise their young on lakes, winter on the ocean, and dive to feed on fish. Large numbers of these various waterbirds gather at the coast for the winter, in freshwater, saltwater, or intermediary habitats, according to species. Between the estuary and the wider area of the Salish Sea, they find their favourite foraging areas and those safest from predators. Here they undergo their "**moult**" in late summer and fall and emerge in winter or early spring with shiny, fresh plumage, ready for the breeding season.

On fine days in late winter, male diving ducks can be seen posturing and displaying to rival males and to females.

Large numbers of geese spend time in winter in the Fraser estuary. They include big flocks of Snow Geese, abundant Canada Geese, and smaller numbers of Black Brant, a sea goose. There is something essentially Canadian about Snow Geese, but those that visit the estuary are in fact from Wrangel Island, off the northeast coast of Russia! They arrive in southern BC in late October, and their large, clamouring flocks gather to feed on roots and rhizomes in marshes and surrounding farm fields. Most move on to the Skagit area of Washington state by early December, returning to feed on the newly sprouting marsh vegetation in late February or March, before making their northward migration. Their numbers have steadily grown over the last two decades, perhaps as milder conditions on Wrangel Island have led to increased breeding success. Brant are the most maritime of geese and stay on the seaward side of the dikes. A flock is often seen at Beach Grove, near Boundary Bay Regional Park. Interestingly, Brant must swallow gravel to facilitate the consumption of eelgrass on which they graze, and are attracted to the gravel bars where a small creek enters the bay.

Swans are another large white waterfowl seen in southern BC in winter; the two species are distinguishable by their calls and bill shapes. Trumpeter Swans recovered from near extinction, and flocks of up to 20 or more may be seen on farmland or flying overhead in "skeins." Their sonorous call notes give them their name. Tundra Swans now occur in far fewer numbers, yet were historically more common. They used to gather in big flocks at Sumas Lake, up the Fraser Valley, but the lake was entirely drained in the 1920s in order to create more farmland for European settlers.

The white cloud of birds on a distant field may not be geese or swans, however. Thousands of overwintering gulls gather to roost on fields, particularly those with little vegetation, such as turf farms or ploughed land. Sometimes they perch in big flocks on farm building roofs. Many of them are Glaucous-winged Gulls, a west coast species that nests on islands off the coast, but other species, confusingly similar in plumage, mingle among them. Gull aficionados will spend time to identify the slightly smaller Iceland Gull, pretty little Mew Gull, or the clearly named Ring-billed Gull. When an eagle flies over, the flocks rise in a flutter of wings and cries, and circle around before descending again.

Some Bald Eagles nest in the delta and many more congregate in winter, attracted by the abundance of waterfowl and other prey. It is not uncommon to see groups of ten or more sitting together on plowed fields or along the shoreline at Boundary Bay in the coldest months of the year. The increase in this species' population over the last 20 years is a remarkable story. It was feared that the birds would be wiped out in much of North America, as their numbers had declined dramatically across their range due to illegal shooting, habitat destruction, and the contamination of their food sources with pesticides. Successful conservation measures, however, led to a huge resurgence in their population, and now many of the delta's large Cottonwood and Douglas Fir trees hold an eagle nest, some of which are occupied by late February.

Many predators gather to eat shrews, voles, and mice that have been forced into the open by rising winter water levels in fields and ditches. The Townsend's Vole, a relatively large and abundant species, is a favoured food for wintering Rough-legged Hawk, Red-tailed Hawk, Short-eared Owl, Great Blue Heron, and Coyotes. Coyotes, members of the dog family, are more often seen out and about during their January–February breeding season.

It was the Fraser delta's natural wealth that compelled me to explore and to dig deeper into its composition, diversity, and history, and then to share this knowledge through books and articles, walks and talks, and speaking up for nature on the many occasions when it was threatened with destruction. My lifelong fascination for birds and nature has led to unforgettable activities and deep friendships around the world, but it is into southwestern BC, especially the Fraser River delta, that I have poured most of my energy. The more one learns about nature, the more one wants to know, and a lifetime is not long enough for all its marvels.

SEEING VAST FLOCKS OF SNOW GEESE ALL TAKE OFF ON A MOMENT'S NOTICE FROM THE FIELDS ON WESTHAM ISLAND IS BOTH A RITE OF PASSAGE FOR BC BIRDWATCHERS AND AN INCREDIBLY MEMORABLE EXPERIENCE. PHOTO ERIC GROSS.

Winter Birds

in Metro Vancouver

AT A GLANCE

WHAT: Birds such as: Barrow's Goldeneye (*Bucephala islandica*), Surf Scoter (*Melanitta perspicillata*), Harlequin Duck (*Histrionicus histrionicus*), Pacific Wren (*Troglodytes pacificus*), Northwestern Crow (*Corvus caurinus*), in various Metro Vancouver regions

WHERE: Fraser River delta / Stanley Park / Mount Seymour / residential neighbourhoods & forests

WHEN: December–March

TRADITIONAL TERRITORY OF METRO VANCOUVER AND FRASER VALLEY: səl̓ ilwətaʔɬ təməxʷ (Tsleil-Waututh), Skwxwú7mesh-ulh Temíx̱ w (Squamish), sq̓əc̓ iy̓aʔɬ təməxʷ (Katzie), Kwantlen, Coast Salish, sc̓əwaθenaʔɬ təməxʷ (Tsawwassen), S'ólh Téméxw (Stó:lō), Á,LEṈENEȻ ŁTE (W̱SÁNEĆ), Semiahmoo, Kwikwetlem, xʷməθkʷəy̓əm (Musqueam), Stz'uminus, Nuxwsa'7aq (Nooksack), Dënéndeh, Qayqayt, Ktunaxa ʔamakʔis, Hul'qumi'num Treaty Group

ACCESS: Easy

In her "Introduction to Winter," Anne Murray shares her affection for the Fraser River delta and the diversity of its wildlife, including birds that migrate through or call the region home. Anne is a long-time volunteer with the BC Nature Important Bird Areas (IBAs) program. At the time of writing, IBAs are being expanded into the network of global **Key Biodiversity Areas** (KBAs) to include all species, not just birds. Key Biodiversity Areas are "among the most incredible and diverse places on Earth for nature, from deserts to the middle of the ocean, [and] are sites of global importance to the planet's overall health and the persistence of biodiversity."[130] The Fraser River delta, long recognized as an important bird area for overwintering and migrating birds, is a critical habitat in

THE BOUNDARY BAY DIKE SPLITS PRIME FORAGING HABITAT FOR NORTHERN HARRIERS. IT'S ESPECIALLY EXCITING TO SEE A HARRIER PASS OVER THE DIKE AT EYE LEVEL AS IT HUNTS IN BOTH THE FORESHORE AND THE FARM FIELDS. PHOTO SABINE DECAMP.

Canada, and is now recognized globally, not only for its birds but for the other species that inhabit it as well. As Anne notes, "In winter, it is the top birdwatching location in Canada!" In this chapter, I would like to pick up on what Anne started in her "Introduction to Winter" and expand the scope to the wider Metro Vancouver region. While reading this chapter, consider how you identify with birds. Do you watch them from your window? Or do you use a feeder to bring them closer? When you walk in a park, do you notice if they're singing? Have you had a special experience with a bird? You see, Metro Vancouver's relatively mild climate makes this region extra special for "snowbirds," at least those of the avian kind!

Birds are often a gateway for getting people, especially children, more closely connected to the environment. Visiting Westham Island's George C. Reifel Migratory Bird Sanctuary (one of BC's most well-known birding hotspots), purchasing a small bag of birdseed, and feeding the Mallard Ducks at your feet or the Black-capped Chickadees that land on your hand is an unforgettable experience (see Engaging and Taking Action). The commotion and declarations of the ducks, or the ten-gram, feathered living being that looks you in the eye while taking an energy-rich sunflower seed from your hand, becomes a memory imprinted deep in the cortex of the brain. Another impressive winter activity is visiting Boundary Bay itself.

WINTER BIRDS IN METRO VANCOUVER

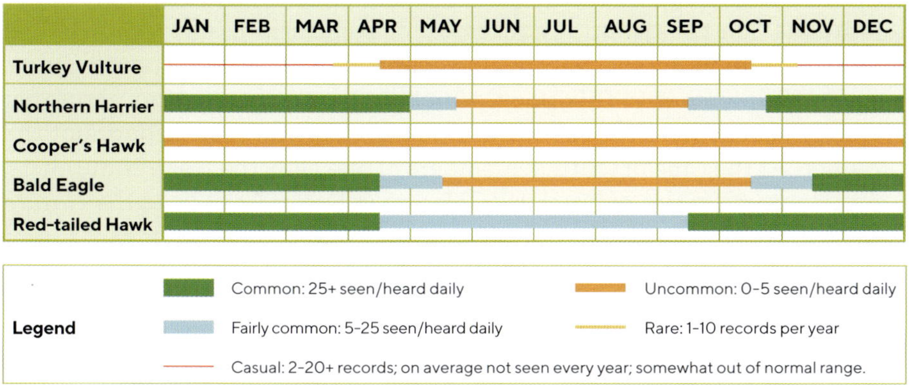

CHECKLISTS SUCH AS THIS ONE HELP US TO UNDERSTAND THE MONTHLY DIVERSITY OF BIRDS IN THE METRO VANCOUVER AREA. SOURCE: NATURE VANCOUVER.

Get yourself to the end of 72nd Street (past the Boundary Bay Airport), arriving at a high tide (because that's when the shorebirds are feeding and lapping up nutrient-rich **biofilm** on the tidal mudflats as the high tide pushes them closer to the shoreline), and cycle or walk the dike, west toward Beach Grove in Tsawwassen or east in the direction of Delta Heritage Park. There, in winter, you will see Bald Eagles, Red-tailed Hawks, Peregrine Falcons, and other raptors, all impressive hunters and powerful apex predators that make birdwatching exciting. My favourite raptor is the Northern Harrier, which, if you're lucky, will glide effortlessly over the dike in front of you, only a couple of metres off the ground, searching for its next meal. In the Canadian winter, there may not be a better place to observe birds of prey than here in the critical wetlands of Boundary Bay. If you opt to venture out for either of these KBA habitat experiences, then be prepared with a pair of binoculars, the Cornell Lab's Merlin Birding App (or a birding ID book), and Nature Vancouver's Seasonal Checklist of Greater Vancouver's Birds. The checklist uses a system of thin to thick bars laid over the months of the year to indicate how abundant birds are at various times of the year. It's very useful!

To make your birding a social experience with the potential to learn from others, check out your local naturalist club at BC Nature. Local naturalist clubs are generally members of the wider Federation of BC Naturalists, and you will find clubs in Vancouver, Delta, the Tri-Cities, Chilliwack, and other BC locations. Content and photos collected from people in local naturalist organizations provided critical contributions to the research and construction of this guidebook, as the members of these clubs are a fantastic "on the ground" source of local natural history knowledge! Members from many BC Nature clubs contribute to the BC Nature Guide, a great complement to this book, which helps the developing naturalist to easily access wildlife and natural features of interest throughout the province. At the time of writing, members of one club, the Delta Naturalists, are designing a series of brochures, *Nature in Delta*. As of 2024 the *Nature in Delta* brochures consisted of animals, bugs, intertidal, birds, and experiencing birds, butterflies and moths, damselflies and dragonflies, mushrooms, and flowering and woodland plants. In fact, many of the images in these brochures came from photographers who also agreed to contribute their photos to this book!

Extending beyond the Fraser River delta, while keeping to the theme of winter birds in Metro Vancouver, an exploration of the Stanley Park Seawall is in order. Many diving ducks utilize the coastal temperate waters off BC's southwest coast, and one location to see seabirds up close is the seawall during the high tides of the winter months. The Barrow's Goldeneye is common; an adult male has a dark bulbous head (even purplish if the sun is shining on it), with a black and white body, and a white crescent on its face. The "golden eye" is clearly visible in both the male and the brown-headed, brownish-bodied female. This is a northwestern landscape bird, breeding inland in remote boreal or interior freshwater lakes but wintering here on the coast.[131] The global population of this bird is approximately 180,000,[132] and southwestern BC is known to be home to a significant number of these birds in the winter. In 1989, the Exxon Valdez oil spill impacted goldeneyes significantly, and a potential future winter shipping spill on the BC coast could have drastic effects on the global population of this species, due to their high wintering numbers on our coast. It's one reason many people don't want more oil tanker traffic on our coast.

A primary source for the global numbers of Barrow's Goldeneye and other long-term bird population trends across North America is the over 50 years of records collected for the annual Breeding Bird Survey. This is an important database used by scientists and wildlife managers to stay informed about significant changes in breeding bird populations. Volunteers have to be skilled birders, able to confidently identify all birds by sight and sound, and be committed to the project for multiple years. It's a global survey, completed in BC on one day only between May 28 and July 7. If you are growing your birding skills, check it out at Birds Canada or Environment and Climate Change Canada by searching "Breeding Bird Survey."

BARROW'S GOLDENEYES ARE COMMON WINTER VISITORS TO THE SOUTHWEST BC COAST. IT IS THEN THAT BC IS HOME TO A SIGNIFICANT PORTION OF THE GLOBAL POPULATION OF THESE BIRDS. PHOTO TERRY CARR.

"Rafts" (large floating groups) of Surf Scoters are another thing to watch for as you walk the Stanley Park Seawall. The male is mostly black, but has a comical orange bill and a white patch on the back of its neck. This is in contrast to the dark-brown female, whose sloping bill is black, with two whitish patches on her face. They nest in shallow wetland lakes, in some of the wildest parts of northern Canada and Alaska. (Deeper lakes have bigger predatory fish like Northern Pike, which snatch their ducklings from below!) In BC, however, we see Surf Scoters as coastal birds that stay close to shore and choose the safety of large flocks while on the water. A favourite place of mine to see this bird in winter is standing on the deck of a BC Ferry, while still in port before a sailing. Look down at the port pilings filled with California Mussels at the terminal and watch the Surf Scoters dive deep to rip them off, popping back up to swallow the shell whole, dissolving it with its strong stomach acids! Much of what we know about the health of seabird populations like the Barrow's Goldeneye and the Surf Scoter is from the British Columbia Coastal Waterbird Survey, running since 1999. The survey data is collected by volunteers to assess annual changes and

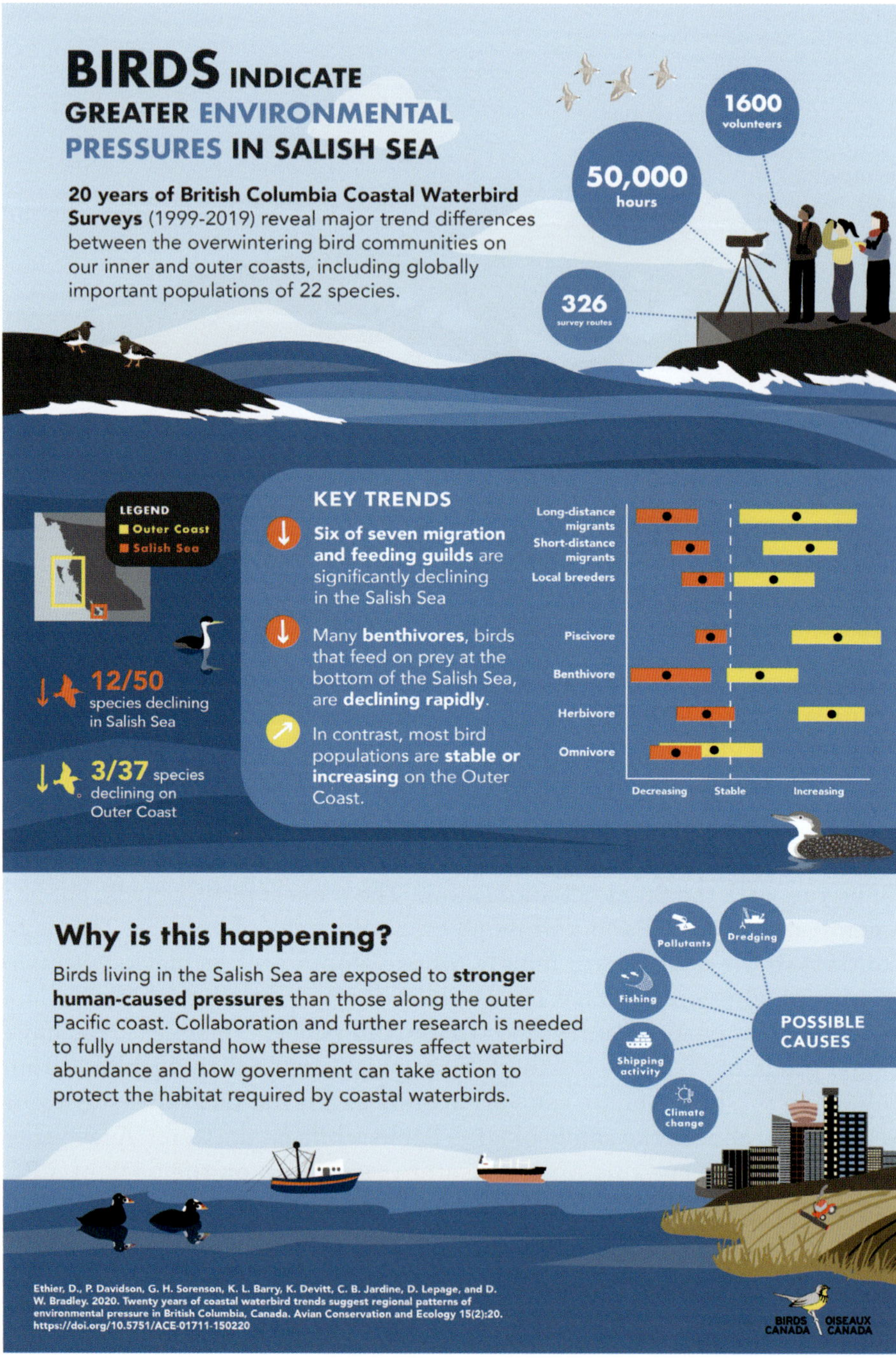

THIS BC COASTAL WATERBIRD SURVEY INFOGRAPHIC, DERIVED FROM 20 YEARS OF SURVEY RECORDS, OUTLINES SOME OF THE KEY ENVIRONMENTAL PRESSURES ON WATERBIRDS IN SOUTHWEST BC. SOURCE: BCCWS DESIGNER, ALICE SUN.

THE MOST WIDESPREAD CORMORANT SPECIES IN NORTH AMERICA, THE DOUBLE-CRESTED CORMORANT, IS LARGER THAN THE PELAGIC CORMORANT AND FLIES WITH A KINK IN ITS NECK. *WATCH FOR THEM DRYING THEIR WINGS BY HOLDING THEM OUT TO THEIR SIDES.* THIS SPECIES MAY ALSO BE SEEN ON LAKES IN THE BC INTERIOR. PHOTO SABINE DECAMP.

long-term trends in BC's coastal waterbird populations and distribution, as well as the effects of human activities on them. The monthly surveys are completed by volunteer birdwatchers with intermediate to expert birding skills. A summary of lessons learned over 20 years of survey collection is outlined in the infographic above.

Among the rafts of scoters around Stanley Park, also look for other coastal waterbirds such as scaups, mergansers, Buffleheads, and Harlequin Ducks. Breeding male Harlequin Ducks are a spectacular slate blue with white stripes and chestnut sides. Their heads are marked with a white crescent in front of the eye and chestnut highlights on the tops of their heads. When you see a Harlequin, it's so stunning you don't forget it! In summer, you are more likely to find Harlequin Ducks in fast-flowing rivers, such as the Similkameen River in Manning Park, but in the winter they are found almost exclusively around exposed headlands such as at Prospect Point, near the Lions Gate Bridge. This is a waterfowl species that is frequently found in the harshest sections of moving water, and because of this they are often discovered to have a history of broken (and healed) bones. With a global population of 170,000, Harlequin Ducks are considered a stable species, but they are another bird that is vulnerable to pollutants and spills. By Second and Third beaches, also look for Oystercatchers, loons, and other seabirds with long necks, such as grebes and cormorants. In the spring Pelagic Cormorants nest on the cliffs at Prospect Point above the waters, where they forage for small fish. Try to distinguish between these and the Double-crested Cormorants that also ply the waters around Burrard Inlet and English Bay.

Moving from waterbirds to urban ones, the forests in Stanley Park are winter homes to Dark-eyed Juncos, Varied Thrushes, American Robins, Spotted Towhees, Black-capped Chickadees, Song Sparrows, and a forest favourite of mine, the Pacific Wren. This very small rufous-brown forest bird with an upturned tail can often be seen in the forested underbrush, disappearing into cavities or crevices as it forages for food. A recent science discovery regarding this bird occurred at UBC's Biodiversity Research Centre, when Dr. Darren Irwin, a professor of zoology, discovered that the spring songs of Winter Wrens (the former name of our coastal species) in coastal BC were different from the songs of Winter Wrens found east of the Rockies. He did his research in BC's Tumbler Ridge, an area where many birds of the east meet birds of the west. Check out where Tumbler Ridge is

NAMED FOR THEIR BOLD PATTERNS, THE MALES OF THIS WATERFOWL LOOK AS IF THEY'VE BEEN PAINTED IN INTRICATE DETAIL. THEY ARE FOUND FORAGING FOR SMALL FISH, MOSTLY IN ROUGH WATER. *LOOK FOR THEM IN THE WINTER AT STANLEY PARK NEAR PROSPECT POINT.* PHOTO TERRY CARR.

on Google Maps, turn on the satellite layer, and you will notice it's located at a *low point* between the southern and northern Rocky Mountains. It's in this low point that many species from central and eastern Canada mix with species from the Pacific Coast, and it was here that Darren noticed differences in the wrens' songs. **Genetic analyses** of the birds' DNA showed they could be divided into two "**genotypes**," corresponding to their different songs. What looked like the same bird (same "**phenotype**") ended up being split into two species – the Winter Wren and the Pacific Wren, based on their different songs, and confirmed by their unique DNA. Most interestingly, the research showed that the last time these two species shared a common ancestor was 4.3 million years ago![133] Pacific Wrens, and the other birds mentioned above, should be able to be located in just about any urban neighbourhood with sufficient forest. Find these areas by moving your satellite-layer Google Map to Metro Vancouver and looking for large green spaces, including Burnaby's Central Park and Burnaby Lake, Vancouver's Pacific Spirit Regional Park, Surrey's Green Timbers Urban Forest, Coquitlam's Mundy Park, or any of the intact forests of the North Shore or the Fraser Valley.

A few additional Metro Vancouver winter bird species worthy of note include Northwestern Crows, Common Ravens, and Anna's Hummingbirds. The Still Creek riparian corridor that parallels the Trans-Canada Highway on the north side at Willingdon Avenue in Burnaby is the site of a notable

A SUNNY MID-FEBRUARY DAY MAY LEAD TO AN EARLY SPRING CALL FROM A PACIFIC WREN, EASILY ONE OF THE MOST DETAILED AND PROLONGED SONGS IN OUR COASTAL FOREST, AND ALWAYS A DELIGHT TO HEAR. PHOTO SABINE DECAMP.

***WATCH FOR RAVENS IN THE MOUNTAINS AROUND METRO VANCOUVER*, IDENTIFIED BY THEIR LARGER SIZE RELATIVE TO A CROW, THEIR BULKY BEAK, AND THEIR WEDGE-SHAPED TAIL IN FLIGHT. THEY MAY SHOW INTEREST IN YOUR LUNCH, *BUT PLEASE DO NOT FEED THEM*.** PHOTO SABINE DECAMP.

METRO VANCOUVER HAS A LARGE NORTHWESTERN CROW ROOST AT STILL CREEK NEAR WILLINGDON AND HIGHWAY 1. ***OUTSIDE OF BREEDING SEASON, WATCH FOR HUNDREDS OF CROWS FLYING OVERHEAD IN THE SAME DIRECTION IN THE HOUR BEFORE SUNSET.*** **IT IS LIKELY THEY WILL BE HEADING FOR THE ROOST.** PHOTO MARION SHIKAZE.

roost for over 10,000 Northwestern Crows. Although car dealerships displaced crows in the 1990s, and the construction of Costco displaced more in 2007, flocks still gather in the trees at dusk to roost each night.[134] Try this: on any evening (with the exception of spring breeding/nesting/young-rearing time when they are more dispersed across the region), watch the sky 30–60 minutes prior to sunset, and when you see a crow flying overhead, note where the intersection of Willingdon and Highway 1 is in relation to you and see if the crow's trajectory is heading that way. It very likely will be!

Now, Northwestern Crows and Common Ravens are similar birds in many people's minds, but they are entirely separate species, and the key to learning to distinguish them is to go where the ravens are and to spend some time with them. In winter, that's usually up to the mountains. A snowshoe to Mount Seymour's Dinky Peak (easy) or Dog Mountain (moderately difficult) are good ways to see this bird. (Check out my *Active Vancouver* book for information on these snowshoeing options.) Although they are

RECENTLY, NORTHWESTERN CROWS HAVE BEEN LUMPED IN WITH THE SPECIES CLASSIFICATION OF THE AMERICAN CROW, BUT MANY WHO LIVE ON THE BC COAST SEE DIFFERENCES IN THEIR CALLS AND BEHAVIOUR, AND QUESTION THE DECISION. PHOTO MARION SHIKAZE.

both from the Corvid family, ravens are much larger than crows (ravens are closer to hawk size), they tend to travel in pairs (crows are seen in pairs too, but much more likely in larger groups), and they have a wedged-shaped tail when flying overhead (crows' tails are straight). As for their calls, a crow makes a "caw," while a raven's call is more like a low guttural croak.

At the other end of the size spectrum from the raven is the tiny (10 cm and 3–6 g) Anna's Hummingbird, which, according to Christmas bird counts, have become much more common as a winter resident in Metro Vancouver over the past several decades. Their increased population is partly owing to residential nectar feeders, non-native flowering plants, and warmer west coast winter temperatures as a result of climate change. If you learn to hear the vocalizations of Anna's Hummingbirds (learn from allaboutbirds.org or the Merlin Bird mobile app), you will soon hear that these birds are far more common in your neighbourhood yards, parks, and residential streets than you realized. Anna's Hummingbirds breed multiple times over the winter, and even start a second nest before the winter babies have fledged from the first nest! Are you interested in breeding and nesting birds?

British Columbia has an online version of a breeding bird atlas. Created by citizen-scientist volunteers and continuously updated, it is the single most comprehensive source of current information on the status of BC's breeding birds. It includes detailed maps that show where species occur and how common they are, and graphs to help the user understand what's behind a bird species pattern of abundance and distribution. Have a look at the Anna's Hummingbird on the atlas website to understand how prolific they are as breeding birds!

As Metro Vancouver moves from winter to spring, the birds begin to move as well. Mark your calendar for the spring equinox (March 20), and begin to watch a real-time online map of birds migrating on Cornell Lab's BirdCast. Based on past and current data, BirdCast provides daily "real-time predictions of bird migrations: when they migrate, where they migrate, and

ANNA'S HUMMINGBIRDS CAN BE FOUND AROUND VANCOUVER ALL WINTER, VISITING HOME FEEDERS AND RAISING *MULTIPLE* BROODS OF YOUNG, ALL BEFORE SPRING HAS SPRUNG! PHOTO CHRISTINE GAIO.

how far they will be flying."[135] The site then adds current weather surveillance radar from across North America to gather information on the numbers, flight directions, and speeds of migratory bird movement. Using BirdCast, along with eBird (an up to the minute, online citizen-science database of bird observations), you can follow (online) the movements of Metro Vancouver's winter birds as they leave, while welcoming spring migrants as they arrive. By early May, Vancouver's spring migrants are celebrated with the annual Vancouver Bird Celebration, a series of events inspired by World Migratory Bird Day.

I'd like to suggest one last idea you might want to adopt in the transition from winter to spring. In his "Introduction to Spring" in this book, Dr. Rob Butler shares the idea of "The Green Wave," explaining, "With each successive day the landscape sheds its winter tawniness for a fresh new verdant green coat...[and within it] there is also a nutrient wave, not readily apparent to our eyes, but not unnoticed by plant-eating insects and birds."

MID-WINTER IS THE TIME TO SET UP FOR RECORDING NATURE'S HAPPENINGS IN A JOURNAL OR ON A CALENDAR. BY MID-FEBRUARY THE DAYS ARE ALREADY GETTING NOTICEABLY LONGER, AND THE FIRST SIGNS AND SOUNDS OF SPRING WILL BEGIN. *START YOUR CALENDAR OR JOURNAL THIS YEAR!* SOURCE: NATURE VANCOUVER.

The winter-spring transition is a time for **phenology**. Sometimes known as "nature's calendar," phenology is the study of biological occurrences, such as migration, breeding, or the leafing out and flowering of plants, in relation to climatic conditions year over year. As winter moves into spring, it's a good time of year to build your nature identification skills. It's a time to watch the shrubs, ferns, and flowers of the forest floor unfold like a million different pieces of origami! A sense of inquiry together with *phenological observations* anchor you to a place, so you become a witness to the way nature responds to change.

Observing nature, and experiencing the associated *pause* in our activities required to do so, can alter our state of mind. It focuses us on the *here and now,* and releases us from the incessant stimulation of the busy human world. Start by choosing a handheld way (device or notebook and pen) to **record in a journal** what's going on in your own backyard or your favourite patch of nature. See if you can notice the seasonal cycles and rhythms of the species you share that space with. *What do you see, hear, smell, or feel on this date?* Record that. *Are there birds, insects, squirrels, fish, or frogs, and if so, what are they doing?* Record that too. *What plants are unfolding and in what ways? What's the day like?* Make a note. *Is the air warm or cool, moist or dry, windy or still? What clouds, if any, are above?* Mention that too. If you sketch, even quick doodles can add to the inquiry. Some people add to their journal every day, some make time once a week, others pick up their notebooks when it occurs to them – usually when they've noticed something particularly special happening. Whatever timeline works for you, having a place to record your observations, set up in advance, will increase your likelihood of continuing with the practice. The really cool part happens next year after you hear your first Swainson's Thrush sing its spring song and you check back to see that last year you heard him sing for the first time on the same day in May!

One way to approach phenology is through a climate change lens because, as climate change proceeds with its uncertain effects, phenological records kept by locals (known as "citizen scientists") will become more and more essential to understanding those effects. Mark L. Hineline writes in *Ground Truth: A Guide to Tracking Climate Change at Home,* "Individual observations of phenophases, even if they are of a single species, when aggregated with the reports of other observers, provide the facts that drive theoretical understanding and make modeling and prediction possible."[136] Keeping notes, if done carefully, can be an act of worthy scientific research. This guidebook outlines many options for uploading your records to citizen science websites such as iNaturalist, eBird, Happywhale, and others.

SOME AMERICAN ROBINS REMAIN LOCAL ALL WINTER, WHILE OTHERS MIGRATE SOUTH TO THE USA SOUTHWEST, MEXICO, AND THE GULF COAST.
PHOTO CHRISTINE GAIO.

SURF SCOTERS SWALLOW SHELLFISH WHOLE, WHERE THEY DISSOLVE IN STRONG STOMACH ACIDS. ***WATCH FOR THEM IN LARGE GROUPS CALLED RAFTS, ESPECIALLY DURING FALL AND SPRING MIGRATION. USE NATURE VANCOUVER'S BIRDS OF VANCOUVER SEASONAL CHECKLIST IN YOUR PLANNING.*** PHOTO TERRY CARR.

Phenology is a great teaching tool for young people to increase their observational skills, and a personal meditation tool for adults. And who knows, it may also release your long-lost childhood *sense of wonder* as you immerse yourself in the natural phenomena around you!

> Large rafts of Surf Scoters and Barrow's Goldeneyes arrive by early November to overwinter on BC's south-coastal waters. Both have digestive systems with powerful acids to dissolve the shells of the mussels they swallow whole! Move inland and you are likely to find Anna's Hummingbirds building their nests in January, raising their young even before spring's Rufous Hummingbirds have returned! All one needs is a camera and some patience to capture the moments they offer up, for those who are willing to see.
>
> —**ROB ALEXANDER,** nature photographer

Engaging and Taking Action

Feeding birds at the Reifel Migratory Bird Sanctuary in the winter is a unique experience. However, it's important to qualify this. Please see the sanctuary's rules for feeding birds. Generally, people should not feed birds in parks or natural areas. The sanctuary chooses seed and grain that will not harm birds, only offers seeds for feeding in the winter, uses the opportunity for education, and creates boundaries to areas where feeding is allowed. As an environmental educator, I have seen the effects of feeding birds: the delight in a child's eyes, the change of attitude in a young adult, and the desire to "give back" from an older adult. Similarly, I have watched an apple core thrown onto the roadside from a car window, with justification that the wildlife will eat it, without considering that the attracted wildlife will also be drawn to a road! I've seen others feeding Cheezies and sandwich bread to satiate the incessant demand of Canada Jays, but not understanding that the Jays are taking and stashing the food for access later in the winter, and that it's then they will find out there's not enough caloric energy in their stash to help them survive a bitterly cold night. The bottom line is, be thoughtful about how you engage with birds, as they already have an incredible number of obstacles that humans put in their way, and you don't want to be adding one more.

Minimizing Your Carbon Footprint

Viewing winter birds in just about any region of Metro Vancouver can easily be done on a combination of bike and transit. Plan your bird observing trip at translink.ca.

Directions and Accessibility

Directions vary due to the choices you make to observe birds. The wonderful thing is, in Metro Vancouver, your choices are almost endless!

Reifel Bird Sanctuary is accessible with gravel trails, Boundary Bay is as well. The Stanley Park Seawall is fully paved and various green spaces around Vancouver (such as Mundy Park in Coquitlam or Central Park in Burnaby) also have paved trails.

Other Locations to View Winter Birds in Metro Vancouver

While learning about birds in Metro Vancouver, take the opportunity throughout the year to connect birds to the habitats they call home. The trails in the old-growth forest next to the Seymour River Fish Hatchery & Education

Centre in the Lower Seymour Conservation Reserve is a very different place than the mid-marsh trail at Minnekhada Regional Park in Coquitlam. The Iona Jetty at Iona Beach Regional Park is a 4 km ocean spit that is different again, offering an opportunity to view seabirds, while the Black Mountain summit trail on Cypress Mountain is at an altitude that is suitable to other bird species not readily found at lower elevations. The City of Vancouver has a Vancouver Parks Map, a great resource for locating neighbourhood green space, including forested trails, as well as ocean and riverfront strolls. The BC Bird Trail highlights sites in the Fraser Valley, South Fraser, and central Vancouver Island. And don't forget Maplewood Flats Conservation Area in North Vancouver! If you are particularly interested in rare birds that might be in the region, check out the British Columbia Rare Bird Alert. Many of these locations are highlighted in my earlier book, *Active Vancouver.*

Additional Resources

1. Anne Murray, who provided the introduction to winter in this book, wrote a wonderful guidebook called ***A Nature Guide to Boundary Bay.***

2. ***Delta Naturalist Nature in Delta*** brochures are available online or at Delta City Hall or other city facilities.

3. **Key Biodiversity Areas** are great places to visit when exploring your local region or globally.

4. ***Ground Truth: A Guide to Tracking Climate Change at Home,*** by Mark L. Hineline. This is a book on how phenology (the study of cyclic and seasonal natural phenomena) can be a way to anchor yourself in a place and witness the way nature is actually responding to change, instead of dwelling on predicted future calamities. Phenology allows you to take an action and share your knowledge with the science community.

5. Contribute to ***HawkCount,*** part of the Hawk Migration Association of North America, which collects hawk count data and keeps a near-real-time database of hawk counts across the continent.

6. BirdWeb: **Seattle Audubon's Guide to the Birds of Washington State** is also very useful for southern BC.

7. Read about the **Seven Simple Actions to Help Birds** as outlined in the Meadowlark Nature Festival chapter of this book.

8. ***Active Vancouver: A Year-Round Guide to Outdoor Recreation in the City's Natural Environments*** by Roy Jantzen is the other book I have written that connects the reader to nature in various popular recreational areas in, and around, Vancouver.

9. The **Living Planet Index** (LPI) is a measure of the state of the world's biological diversity based on population trends of vertebrate species from terrestrial, freshwater, and marine habitats. There is a Canadian report as well.

10. **Coastal Squeeze** – The coastal birds of Metro Vancouver are not the only ones that feel the loss of habitat; residents of the region will increasingly feel it too in the form of a coastal squeeze. The City of Vancouver highlights sea level rise in their climate change adaptation strategy, and *The Tyee* provides a good summary of the concept in a spring 2021 article suggesting, "When sea levels rise in a natural environment, a marsh will migrate inland, staying close to the intertidal zone. But a problem arises in coastal cities when we build infrastructure to protect human development from flooding: the marsh is not able to migrate, and drowns when the water gets too deep."[137]

CLIMATE CHANGE AND METRO VANCOUVER'S WINTER BIRDS

Consequences of climate change will likely be mixed for Metro Vancouver's birds. Warmer winter temperatures will shift the ranges of birds, such as we have seen with Anna's Hummingbirds. Those birds that are very adaptable, like the Northwestern Crow or Common Raven, will likely be more resilient than specialist birds such as Peregrine Falcons, Northern Harriers, or Harlequin Ducks. One suggestion in this chapter is to record phenological observations around your own home to help you notice and understand the uncertain effects that will result as climate change proceeds. A second suggestion would be to support wetlands and connect to World Wetlands Day.

BETWEEN THEIR FOURTH AND FIFTH YEARS, BOTH MALE AND FEMALE BALD EAGLES GET THE REGAL LOOK OF THE WHITE FEATHERS ON THEIR HEADS.
PHOTO ROB ALEXANDER.

Bald Eagles

at the Chehalis Flats Bald Eagle & Salmon Preserve (Harrison Mills) & at Brackendale (Squamish)

AT A GLANCE

WHAT: Bald Eagles (*Haliaeetus leucocephalus*) feeding on salmon
WHERE: Harrison Mills at the Chehalis Flats Bald Eagle & Salmon Preserve and Brackendale
WHEN: Third week of November for Harrison Mills, and first week of January in Brackendale
TRADITIONAL TERRITORY: Stz'uminus, S'ólh Téméxw (Stó:lō), Coast Salish and Sts'ailes at Harrison, Skwxwú7mesh-ulh Temíx̱w (Squamish), Coast Salish at Brackendale
ACCESS: Viewing platforms in Harrison and dike walking in Brackendale

Try to think like a BC eagle for a moment, a migratory Bald Eagle. Your summers are spent in northern BC, Yukon, and Alaska, on lakes and rivers abundant with Lake Trout, Northern Pike, Arctic Grayling, and Whitefish. As autumn proceeds, northern lakes and rivers freeze over and the salmon carcasses you've been feeding on disappear under the snow and ice. You need to move south. Where to go? As an opportunistic feeding aquatic bird, it turns out you're most likely to move in the direction of the British Columbia coast, congregating with many of your compatriots on the southwest coast and the Fraser Valley, arriving in November or December. You can watch what this migration looks like through data gathered on eBird's Status and Trends Abundance Animation. (Be sure to slow it down and watch it frame by frame to get a detailed sense of the migration movement.) Long-time Bald Eagle conservationist David Hancock shared his thoughts with me about this well-known predator, and

THE DIKE IN BRACKENDALE OFFERS A FULLY ACCESSIBLE VIEWING AREA FOR ANYONE TO WATCH BALD EAGLES, ESPECIALLY IN FALL DURING SALMON SPAWNING. IF YOU TIME IT RIGHT, THERE WILL BE VOLUNTEERS THERE WITH SPOTTING SCOPES AND A WEALTH OF KNOWLEDGE TO SHARE. PHOTO JUDITH KNAPP.

he estimates that 35,000 of these migrants pass through the Fraser Valley on their southern migration, with 10,000–15,000 of them settling around the Harrison Mills area at the Chehalis Flats Bald Eagle & Salmon Preserve. This is the world's largest gathering of Bald Eagles! David told me that it's the combination of weather conditions and diminishing salmon returns in northern rivers that is the biggest driver promoting eagles to migrate south to the Chehalis and Brackendale areas.

Geographically, the Chehalis is a large alluvial fan (an accumulation of sediments), with a myriad of rivulets (small streams) where it meets the Harrison River. Northwest of the Chehalis, another approximately 1,500 Bald Eagles tend to congregate in the Squamish-Brackendale area (Brackendale Eagles Provincial Park), which is also near the confluence of rivers, specifically the Squamish, Cheakamus, and Mamquam rivers. In addition to these migrant eagles, our local Bald Eagles, which nest and breed in Metro Vancouver, the Fraser Valley, and southern Vancouver Island, use this area as well. You can see their nests in many of the big trees surrounding the city and its suburbs. As of 2021, there are 21 nests in Vancouver, 43 in Surrey, and 83 in Delta. In fact, as of May 2024, there are 565 known active nests in the lower Fraser Valley alone. But it hasn't always been this way.

David Hancock started his surveys in 1953, and he recorded only three nesting pairs in 1963. At that time, eagle numbers were low and attributable to a bounty placed on them (money distributed for each Bald Eagle killed), implemented because it was thought they were causing the decline of salmon stocks. Research later showed that the decline was in fact caused by other factors, including overharvest in the commercial fishery. Additionally, **DDT** (Dichlorodiphenyltrichloroethane), a popular pesticide used in North America in the 1960s, also had enormous impacts on eagles. As the pesticide bioaccumulated up the food chain to this bird of prey (see the explanation of "bioaccumulation" in the chapter on Turkey Vultures), it softened the shells of eagle eggs, undermining their ability to hatch properly.[138] Since DDT was banned by the end of 1990, the Bald Eagle has gone from the brink of extinction within the contiguous USA to a not-so-uncommon bird to see today. Therefore, the Bald Eagle, the American national symbol, is known widely as a conservation success story, and those who wish to see them in all their glory know their greatest concentrations are in Canada, specifically southwestern BC.

The protection of both the Chehalis Flats and the Brackendale Bald Eagle habitats can be attributed to the community support that surrounds them. Locals in both locations have supported habitat protection, eagle counts, education, and festivals to celebrate the annual return of the Bald Eagles (approximately the third week of November for Harrison/Chehalis and first week of January in Squamish/Brackendale). In addition to protected spaces, the eagles also require low water levels (high water flushes salmon carcasses down the river, while low water ensures salmon carcasses are scattered about the streambanks); abundant salmon (though there are serious concerns about many coastal BC salmon populations); and protection from disturbance (walkers, kayakers, paddleboarders, sport fishers, photographers, campers, and other recreationalists). *Please keep off the flats from October to February* as that's when the eagles feed, rest, and socialize at the river's edge, and "flushing" them (making them fly away in alarm) leads to the unnecessary burning of calories. The use of drones for photography is also highly disturbing for the eagles.

The Sts'ailes (pronounced sta-ay-liss and formerly known as the Chehalis First Nation) name for the Chehalis Flats Bald Eagle & Salmon Preserve is Lhá:lt, and it is now a **Wildlife Management Area** (WMA). WMAs are lands designated under the BC Wildlife Act to protect regional or internationally significant fish and wildlife species, or their habitats. The Chehalis Flats are

IN THE WINTER, THE PRESENCE OF EAGLES AT BOTH CHEHALIS FLATS AND BRACKENDALE IS DUE TO THE DEAD AND DYING SALMON IN THE SHALLOW EDGES OF THE RIVERS. IN GOOD SALMON YEARS, IT'S NOT UNCOMMON FOR EAGLES TO FEED PREFERENTIALLY ON THE HIGH-NUTRIENT EYES, BRAIN, AND EGGS. PHOTO TIM CYR.

known as a salmon stronghold and host one of the last consistently large runs of salmon in North America. The Bald Eagles learned this long ago! All five species of BC salmon spawn in the Chehalis River, along with Steelhead, Rainbow, and Cutthroat Trout.[139] The Pretty Estates eagle viewing platform at Sandpiper Resort is among the best places to watch eagles from November to January, with eagle numbers often peaking around mid-December. A second location is the Eagle Point Park observatory about one kilometre upriver. Near Squamish, the prime viewing location in Brackendale is the Eagle Run dike and viewing shelter, which includes an interpretive display. It is opposite the BC Easter Seals camp on Government Road and directly across the Squamish River from Brackendale Eagles Provincial Park. Bald Eagles are creatures of the salmon, and for them to prosper, salmon must thrive as well.

IT'S IMPORTANT TO KEEP TO THE VIEWING TOWER AT CHEHALIS FLATS AS IT DOESN'T TAKE MUCH TO SPOOK THE EAGLES FROM THE SIDE OF THE RIVER. ***THIS MAP SHOWS THE NO-GO ZONE ON THE FLATS WHERE THE EAGLES FEED.*** SOURCE DAVID HANCOCK, HANCOCK WILDLIFE FOUNDATION.

As a professor and naturalist, I often hear recurring questions from students about their observations of the natural world. In a forest, questions are usually based around the ages and types of trees. With Bald Eagles, it's "Do they mate for life?" and "Why do some have white heads and others don't?" The Hancock Wildlife Foundation has a fantastic website that covers many common (and not-so-common) questions, with input from David Hancock, the foundation's founder, and a devoted student of / expert on Bald Eagle behaviour. For example, Bald Eagle pairs are known to mate for life, but they don't spend all year together. They arrive separately in southwestern BC around October, having left their nest sites and fledgling young the previous July to journey north. The pair reconfirms their bond with each other when they return in the autumn, with behaviours such as matching head and body movements and calling in unison. It's at this time of year you may see the couple flying in tight unison with one bird slightly behind and above the other. The behaviour is known as **pair bonding** and is spectacular to see. They'll likely return to the same nest they previously used and begin to restake their claim to their previous territory. The timings of their return

THE OBSERVATION DECK FOR BALD EAGLE VIEWING AT CHEHALIS FLATS NEAR HARRISON MILLS IS BROAD ENOUGH TO SHARE WITH THE WHOLE FAMILY OR IN SOLITUDE.

from their northern migration, the re-establishment of their nests, the laying of their eggs, and the hatching of their young are extremely consistent, year after year. (In fact, if you want to be a citizen scientist, you can observe this for yourself by watching and recording the daily happenings on one of the Hancock Wildlife Foundation livestream eagle cams, and then repeating the process again the following year, and the year after that.)

Bald Eagle nest building or repair happens in earnest in January, while mating can continue from their autumn rebonding through until March. Females lay one to three eggs in mid-March, spaced about three days apart, with hatching happening 36 to 37 days later, and therefore chicks also hatch three days apart. If you'd like to watch the hatching, an incredible natural event to witness on the "**nest cams**," mark your calendar for the first to third weeks of April. (According to David, hatching often happens overnight or early in the morning, so it's probably best to choose a nest cam that has an infrared camera!) Hatchlings are born with huge feet to hold up their equally huge heads and beaks; the feet are necessary to balance themselves and to fight their siblings for food brought to the nest from parents. The nest

BALD EAGLES ARE VERY TERRITORIAL AND IT'S NOT UNCOMMON TO NOTICE A MATURE EAGLE PRESSURING A JUVENILE FROM A PRIME TREE BRANCH OR CHASING IT OUT OF ITS FEEDING AREA. PHOTO MICHAEL SCHMIDT PHOTOGRAPHY.

cams, by the way, aren't for the squeamish, as eagles are predators and the food delivered is their prey. Nonetheless, it's great fun to watch the eaglets over a period of months, as they grow greyish down feathers (they are born covered in white down), form solid wings from stubby wing butts, and exercise their new-found agility through physical antics on the nest or nearby branches. By the 12th week, the juvenile eagles "fledge," leaving the nest on their own and they remain in their birth area. The parents stick around for another four to six days, still hunting for and feeding their young, but they *never train them to hunt*. Instead, they leave them to figure that out on their own. By then it's July and the parents go their separate ways north to feed, rest, and recuperate from the work of parenting, getting themselves ready to meet each other again in about 12 weeks' time to do it all again!

The Bald Eagle pair's breeding period spans almost ten months, beginning with pair bonding, then territory occupancy, territory defence, nest repair and building, egg laying, incubation, feeding of their young through the fledgling and post-fledgling phases, and then finally abandoning their territory and their young as they part ways and head northward. It's important to reiterate that the success of *all* of this is *dependent upon the timing and abundance of spawning salmon*. Having access to an abundance of salmon carcasses is especially important for the newly fledged juveniles, as

they're almost completely unable to catch live prey during their first year, and it takes several more years for them to become experienced hunters and fishers. Maturing skills seem to mirror their maturing appearance, as they don't get their fully white head and arrive at sexual maturity until their fifth year, and generally they don't begin to breed until their sixth year.[140] This suggests it takes a long time to acquire the experience necessary to hunt effectively and to defend a territory, skills necessary to provide for and protect their young. For those interested in learning more, resources such as the breeding bird atlas and Breeding Bird Survey can help take a wildlife viewer

MATING PAIRS ARE FUN TO WATCH IN THE WILD OR ONLINE USING AN EAGLE CAM. GENERALLY, WHEN A PAIR IS SEEN SIDE BY SIDE, THE FEMALE IS THE LARGER OF THE PAIR, WHICH IS COMMON FOR BIRDS OF PREY.
PHOTO ROB ALEXANDER.

from being a casual observer into a deeper realm of ecological literacy. (See the chapter on Winter Birds at Boundary Bay for a deeper understanding of breeding birds in our area.)

Greater knowledge often leads to more intense questioning about a species, so before long, interest about size, speed, or lifespan leads to deeper thoughts about a species' role in its greater ecosystem. There is always so much more to learn! For example, Bald Eagle females are approximately 25 per cent larger than males. (In the world of species this difference in a characteristic such as size is known as "**sexual dimorphism**".) Male or female, their size is impressive and generally just under a metre in body length, with a 2 m wingspan, and 4–6 kg in weight. They fly and glide at about 50 km/h when overhead, while their hunting dive speed can be as great as 150 km/h. But the thing I love the most is watching them through binoculars as they catch a ride on a rising column of warm air, known as a "thermal." It takes my breath away every time! If Bald Eagles manage to stay healthy, they live an average of 20 years, and during that time their role in their ecosystem is substantial. Just consider one aspect of their niche – how they transfer marine-derived nutrients from sea to land in BC coastal ecosystems through their feeding behaviours.[141]

A Bald Eagle call is a high-pitched staccato of chirping whistles, somewhat similar to that of a gull. When I hear the distinctive call, it's a special moment and I find it elicits an emotional connection for me to the place I'm in when listening. If just hearing the call of this magnificent bird doesn't do that for you yet, then consider this. Fossil records show that the Bald Eagle diverged evolutionarily from the White-tailed Eagle of Eurasia (its closest relative) sometime between 10 and 28 million years before the present day, meaning it has inhabited this land for millions of years! (See Appendix C for more on this idea.) So perhaps their high-pitched whistling or piping notes are more than mere calls; perhaps they are announcing to the world their ancestry and their knowledge about the depths of time during which they've frequented these Pacific skies.

> Bald Eagles, rare and wild in the 1950s, are today breeding in huge numbers, throughout Metro Vancouver and its three million humans that share this wondrous corner of Super Natural British Columbia. In my studies of the last 65 years I have seen the entire Metro Vancouver eagle population go from three breeding pairs in 1964 to more than 560 pairs today. And our Lower Fraser Valley is also the winter home of more than 35,000 northern Canada and Alaska breeding eagles. This is quite wonderful if you are a naturalist!
>
> —**DAVID HANCOCK,** Bald Eagle researcher, Hancock Wildlife Foundation

Engaging and Taking Action

Most Metro Vancouver communities have Bald Eagle nests. Is there one that you know of in your community? If so, consider visiting the Hancock Wildlife Foundation volunteer page and sign up to help increase our understanding of either the success of breeding eagles, or how they're adapting to human activities in their nest areas. As a nest monitor, you might even gather scientific data that could be used to make it more difficult for developers or landowners to cut down the mature trees eagles have nests in (believe it or not, that really happens!), or to prevent other human activities that are detrimental to eagles. Another way to take action is by volunteering to be a nest "cam" monitor, which would allow you to help directly from your computer (you can even control the cameras to focus on what the eagles are doing!). Online volunteers make it possible to have a more thorough understanding of the activities that transpire at the 40 or so local eagle nests that have cameras focused on them.

Engaging can also be as simple as donating through an adopt-a-nest program and then following the life cycle, behaviours, and concerns affecting your particular eagle, and sharing this knowledge with others. Another option is to participate in an annual eagle count, one of which takes place in mid-November at Harrison, and another in early January in Brackendale. To participate, contact the Hancock Wildlife Foundation for Harrison, or EagleWatch for Brackendale. Educational and festival event

EAGLEWATCH HOSTS SCHOOL AND PRIVATE GROUPS FOR INTERPRETIVE PROGRAMS IN BRACKENDALE. *BEFORE YOUR VISIT, CHECK OUT THEIR WEBSITE TO SEE WHAT THEY HAVE PLANNED.* PHOTO WOLFGANG PAULI.

opportunities also exist in each location. It's important to know that these are *community-led* initiatives that wouldn't happen without the engagement of dedicated volunteers. For the general public, community groups, and schools, they include educational programs, outreach events, and festivals, and even opportunities on weekends to view eagles through spotting scopes, read interpretive panels, and interact with educators.[142]

In addition, you could help with Hancock Wildlife Foundation projects by contributing to their Bald Eagle nest "tree bank" or by helping to garner support for habitat conservation through the creation of Bald Eagle preserves. Do you have a property with large trees suitable for eagle nests? The Hancock Foundation offers a very unique way for property owners with such trees to take action. They've built dozens of artificial nests for Bald Eagles on private property to help increase available nesting habitat. The cost to the homeowner is about $3,500 (2024 price), but if you're interested, please contact David Hancock at the foundation.

Minimizing Your Carbon Footprint

Have you considered renting an electric car for a day to try out what it would be like to travel without being propelled by fossil fuels? I can think of two stunning drives you can do with a one-day rental. The Sea to Sky Highway

IT'S NOT UNCOMMON TO SEE DOZENS OF EAGLES RESTING IN TREES NEXT TO THE RIVER ON A WINTER'S DAY. PEOPLE TRAVEL FROM AROUND THE WORLD TO SEE AN EVENT LIKE THIS, ONE THAT VANCOUVERITES HAVE VIRTUALLY IN THEIR BACKYARD.
PHOTO CHRISTINE GAIO.

to Brackendale, and Highway 7 through Mission, Dewdney, Deroche, and Harrison Mills for Bald Eagle viewing, and then on to Harrison Hot Springs. Pair your eagle viewing experience with learning about the BC Hydro, ChargePoint, or Tesla charging networks. (Say goodbye to gas and hello to class!) For a local Tesla rental, check out Turo.com or contact the Vancouver Electric Vehicle Association (veva.ca) for their suggestions on this. Alternately, you can register for and find a local Evo for the day, supplied by the BC Automobile Association car sharing service.

Directions and Accessibility

From Metro Vancouver you can travel on Highway 99 to Brackendale or Highway 7 to Harrison Mills for Bald Eagle viewing.

The Brackendale dike has excellent accessibility. The viewing platform is accessible to wheelchairs, portable toilets are on site, and benches offer places to rest while walking on the dike. The Chehalis platform also is accessible, though it does require travelling for about half a kilometre on a gravel trail through the golf course at the Sandpiper Resort.

Other Locations to View Bald Eagles

Bald Eagles prefer high perches, so if you can place yourself in a coastal area with large trees near the ocean or rivers, your chances of seeing one are good. Get to know their calls before going out to see them. Even with their great size, it's not unlikely you'll hear them before you see them as they soar above your head. Scan the tops of large dead trees or bare tree branches. Lighthouse Park, Pacific Spirit Regional Park, Capilano Regional Park, Porteau Cove Provincial Park, Boundary Bay, or Delta near the landfill, are all great coastal locations to view eagles in Metro Vancouver.

Other Area Interests

1. Ten kilometres east of Mission is Inch Creek Hatchery in Dewdney. The hatchery is a host site for the annual Fraser Valley Bald Eagle Festival. Many displays explaining salmon and wildlife are available during this event. The hatchery is also home to two adult captive male White Sturgeon in their tanks, which Vancouver Island University uses for their sturgeon breeding program. If you've never seen this ancient species before, I highly recommend a visit.

BALD EAGLES FEED FROM THE RIVER, MAKING IT ESPECIALLY IMPORTANT TO WATCH THEM FROM AFAR, ENSURING THEY DON'T FEEL THREATENED AND FLY AWAY FROM THEIR MEAL. PHOTO TIM CYR.

2. Although October is a little early for Bald Eagles returning to the coast, if you go to the Weaver Creek Spawning Channel around October 15–20, you will likely see Sockeye Salmon at peak spawning time.

3. Xá:ytem (also known as Hatzic Rock) in Mission is a Canadian National Historic Site. This archeological site shows Indigenous occupation on the northern bank of the Fraser River from as much as 5,000 years ago. It has great spiritual value to the Sto:lo, the local First Nation that calls this area home.

4. Have you ever seen an American Dipper? It's a special day when I see one. The dipper is a tiny songbird that lives around rushing, unpolluted salmon streams. It dives underwater to feed on insects, their larvae, salmon eggs, and small fish. You will most likely see them in the fall and, if you're lucky, you will hear their melodious song.

Additional Resources

1. Check out the ***BC Breeding Bird Atlas*** and Cornell Lab / Birds Canada Breeding Bird Survey.

2. The **Hancock Wildlife Foundation** has livestream eagle cams. Get ready to watch these in April!

3. The Squamish Environment Society has an **EagleWatch** Interpretive Program.

4. **eBird's Status and Trends** provides high-resolution data, visualizations, and tools describing where bird populations occur and how they change through time. It is powered by eBird data and updated annually, providing you with the best available science.

CLIMATE CHANGE AND BALD EAGLES

Bald Eagles have experienced challenges, such as DDT poisoning, in the past and their numbers have rebounded, showing they are resilient. However, a key food of theirs is salmon, and if salmon numbers continue to decrease, this will have a cascading effect on many species, including Bald Eagles. Eagles in Metro Vancouver have shown an interest in other food sources, including Metro Vancouver's landfill, and therefore it's possible climate change will lead to more Bald Eagles in urban spaces, foraging for leftover human scraps – but much of this waste product is now being placed in "green boxes" and is now lost to the eagles.

MOOSE, LIKE THIS FEMALE, REQUIRE MANY CALORIES TO SUPPORT THEIR LARGE BODIES. THEREFORE, IF YOU SEE THEM, IT'S LIKELY THEY WILL BE FORAGING AND FEEDING WHETHER ON FRESHWATER PLANTS OR THE BARK OF SHRUBS. PHOTO ERIC GROSS.

Moose

on Green Mountain at Wells Gray Provincial Park

AT A GLANCE

WHAT: Moose (*Alces alces*)
WHERE: Hemp Creek, shortly before Green Mountain Road in Wells Gray Provincial Park
WHEN: Mid-February to mid-March
TRADITIONAL TERRITORY: Secwepemcúl'ecw (Secwépemc), Dakeł Keyoh (ᑕᗸᒡ ᗱᘇᐦ), Dënéndeh, Ktunaxa ʔamakʔis
ACCESS: Moderate (due to distance and commuting in the winter)

Standing about two metres tall at their shoulder, with broad, parabolic-like antlers spanning slightly less than their height, and weighing in at about the mass of a family camping trailer that roams the park roads in the summer, a bull Moose is a formidable sight! Moose are the largest and heaviest members of the deer family, and three subspecies of Moose inhabit much of British Columbia's boreal, evergreen, and temperate-broadleaf forests. There are somewhere between 110,000 and 185,000 Moose in BC, a significant proportion of the slightly more than one million living in the northern regions of North America (which includes most Canadian provinces plus, Alaska and as far south as Utah).[143] So, if you live in southwestern BC, where would be the nearest, most accessible place to reliably find a Moose in winter?

A good answer is the habitat of Wells Gray Provincial Park, which provides valley bottoms with nutritious shrubs and trees for browsing right in the depth of winter snow, along with the high elevation plateaus necessary for summer foraging. Add in other habitat requirements, like lakeshores, salt licks, beaver ponds, and wetlands that Moose use for feeding on aquatic plants and cooling themselves in summer, as well as the logged areas (outside of the park boundaries) or burned areas (both outside and inside the

park) that promote fresh forage opportunities, and you have prime Moose habitat! To meet the energy requirements of such a large body, Moose need to eat about 20 kg of food every day in the winter. So, by early fall, the Moose migrate from summer's higher plateaus to the lower valley bottoms to feed on nutrient-rich, high-sugar-content shrubs such as willows, Red Osier Dogwood, and Highbush Cranberry, as well as, but to a much lesser extent, trees including Black Cottonwood, Paper Birch, Aspen, and Subalpine Fir, making the park a prime location for winter viewing.

The east-facing slopes of Green Mountain are among the most reliable locations to see the park's Moose in winter. Park your vehicle at a pullout 1800 m north of the Flat Iron Trailhead (before Corral road on your right), and 30.6 km from the Clearwater visitor centre. Bring good binoculars or, better yet, a spotting scope. Also watch for signs of Moose, such as tracks, scat,

A MOOSE PAIR IS A RARE SITE, SEEN MOSTLY DURING SPRING BREEDING SEASON. OTHERWISE THEY ARE MAINLY SOLITARY ANIMALS. PHOTO RYAN HAGERTY/USFWS, "ALASKAN MOOSE PAIR," FLICKR, CC BY 2.0.

resting areas, or low shrubbery that has been browsed. Be prepared by knowing how to read these signs before you go. The other potential viewing locations are from the ski trails at Majerus Farm (41.5 km north of the visitor centre on Clearwater Lake Road), or, if you're lucky, you'll come upon them at the roadside. As a note of caution, Moose on the roads present significant danger of impact for moving vehicles, especially when you consider the length of their legs and the placement of their body mass relative to a vehicle's windshield. So use extra caution in Moose territory, particularly when driving at dawn and dusk – the times Moose are most active.

One of the things I love most about natural history is how plants and animals adapt to their environments, and those that live in the harshest environments often have the most amazingly adaptive solutions! Moose are among the most cold-adapted mammals in Canada. Their exceptionally thick hides and two layers of dark brown to blackish fur, with long, hollow, insulating guard hairs and a soft, woolly undercoat, keep Moose warm in winter. But in summer, this becomes a challenge of heat tolerance. So when it's hot, they seek out shaded forests, cool breezes, and above all, immersion in cool water. Hence, they can often be found in the summer wading or swimming in lakes or ponds. Fortunately, their thick fur and hollow guard hairs also help with flotation, making them excellent swimmers, so they can often be found feeding on aquatic plants as they go. They are even known to dive up to 5.5 m to reach plant material on lake bottoms! Moose hooves are another adaptation with multiple purposes. They splay under the animal, increasing its surface area on the snow (like snowshoes) and increasing its swimming efficiency (like flippers). Its long nose ("proboscis"), which likely evolved as an adaptation to aquatic browsing, gives it a keen sense of smell to locate sources of water, food under the snow, mates, or nearby predators.[144] Male Moose are known as "bulls," and it could be said that their ungainly antlers appear to be an exception to these other tailor-made environmental adaptations. They use their antlers in posturing among other males to show their competitors and the females their health in autumn during "the rut." It is then that antlers prove to be worth their weight!

Moose are solitary animals that don't form social bonds like other species of ungulates (hoofed mammals such as Bighorn Sheep or Elk) do. But, in the fall, bulls respond to a female "cow's" long, quavering moans – audible for up to 3 km! This coincides with the buildup of high levels of testosterone in the males, as their bodies prepare for mating. It's important to note that cows have some control over the selection of their mate, and body size (approximately 400 kg), antler size (up to 1.8 m on a prime bull), and antler symmetry (which reflects health) are key determinants. It has been shown that cows "protest moan" more when being courted by smaller, less attractive males, thereby announcing their presence to other nearby males, who then show up for a fight. The more she moans, the more likely a bigger, more attractive Moose will show up, and the more likely she will benefit from the battle, which often entails more posturing than actual fighting. There are two times of year when Moose become more aggressive and unpredictable around people – during the full rut for bulls and in the spring when cows have young near them. In fact, fall is the most dangerous time to encounter a Moose, but they aren't pleased at any time of year if you surprise them. Even though Moose generally give warning signs through their body language before they charge – eye contact, laid-back ears, and just before a charge the hairs on its neck and shoulders stand up – humans are not the best at reading those signs, so it's best to give them lots of space!

After mating, there's an eight-month gestation period, with a calf born (usually one, but twins are not uncommon when food is plentiful) in late May or early June. The social bond between cow and calf is really the only strong bond between Moose, and even this bond is tenuous after as little as a year, when mothers are known to chase off their young, forcing them to fend for themselves, including fending off predators. Separating from her calf allows a mother who is in good condition to focus on pregnancy again the following year, but it's not always good for her calf. A 20-year study from 1996 to 2015 by the government of BC presented calf mortality due to predation as the main contributing factor to steep declines (29 per cent) in BC's Moose population during that same period.[145] What they haven't been able to determine yet is what has caused the increase in predation.

The opening of roads into landscapes, especially roads in the northern boreal range of Moose habitat, may be a factor in the predation and decline

MOOSE ARE WELL SUITED TO A BOREAL WINTER ENVIRONMENT. THIS MEANS THAT WARM SUMMER DAYS REQUIRE PLACES TO COOL OFF, AND ARE OFTEN A TIME YOU SEE MOOSE IN THE WATER, LIKE THIS BULL MALE. PHOTO GOVERNMENT OF ALBERTA, FLICKR, CC BY-NC-ND 2.0.

of Moose in BC. Roads lead to increasing human land settlement and resource development. For instance, the Site C Dam, near Fort St. John in northeastern British Columbia, will flood a large region, much of which is prime Moose habitat. Roads also provide easier access for the movement of some of the main predators of Moose calves, including Grey Wolves, Black and Grizzly Bears, and Cougars. That said, Moose between the ages of 2 and 8 years old are seldom killed by wild predators, but roads also provide access for human poachers, who are less likely to be restricted by the age or size of their quarry. Interestingly, another issue with road development is that it provides access for White-tailed Deer to expand into new territories. Although on the surface deer wouldn't appear to be a major competitor to Moose, in fact, deer carry liver flukes and brain worms, to which the Moose have not adapted a natural defence. When the deer pass those parasites on to unprotected Moose, it can weaken the Moose, causing populations to decline. Moose habitat enhancement in BC may include the "deactivation" of forest service roads and the planting of fast-growing species in riparian

THE WETLAND HABITAT OF THE WIDE CLEARWATER RIVER VALLEY PROVIDES MUCH PRIME MOOSE TERRITORY. PHOTO FRITZ SCHAER.

areas, but Moose habitat covers an incredibly expansive area and there are limited funds to pay for habitat enhancement.[146]

Moose select habitat based on the trade-off between predation, food availability, and snow depth. With climate change, Moose, including those in Wells Gray Park, are more prone to environmental stressors such as warmer temperatures. The BC government states that the provincial average temperature has increased approximately 1.4°C between 1900 and 2013.[147] This not only affects snow depths but also makes Moose more susceptible to introduced pathogens, such as those mentioned from deer. When Moose become heavily infested by ticks, for example, they lose a significant amount of blood to the ticks, and the itching can lead to them rubbing their fur coats down to raw bare skin, opening them up to infections and other challenges, especially in the winter.

In rural Canada, people have relied on healthy populations of Moose, but Moose have not always existed in southern BC. They migrated south as the last ice age receded, and arrived in the Wells Gray region in the mid-1920s, following the growth of young willows and other shrubs after a major fire. In fact, Wells Gray Provincial Park was established in 1939 for reasons other than what we usually create parks for today; it was established as a park for hunters of Moose, and anglers to enjoy the fishing. Moose in other regions of BC *did* figure significantly in First Nations culture long before the early 20th century, however. They were hunted for their meat, their hides were important for shelters, leather moccasins, and clothing, and the bones and antlers were made into a myriad of tools.[148] Today, many subsistence hunters are drawn to pursue Moose for their meat, their antler racks, and their hides, and in BC there is a provincial "harvest," with 8,000–10,000 tags available for hunters.[149]

Like other animals that are hunted in BC, Moose are a "managed" species. That is, *our* decisions and preferences influence their populations. Government documents refer to those who have an interest in Moose as "user groups" or "stakeholders." But there are other "users" of Moose besides us. Just ask the wolves, bears, and Cougars! People typically hold an opinion about wild animal management and how we share space with other species. I believe it is good practice to try to understand both *management and ecological principles* more thoroughly before settling on a firm conviction. Management of a species is complex, and past attempts at it in Canada have often had devastating results: the east coast Cod fishery and the west coast Pacific Herring fishery collapses (see chapter on Pacific Herring) resulted from historically high harvest management quotas for these species. The Northern Spotted Owl in BC has lost most of the old-growth habitat it requires, which, along with other factors such as an increase in Barred Owls, has decreased their numbers to less than a dozen Northern Spotted Owls left in the province![150] In the Wells Gray region, one of the most significant wildlife management issues is that of the Southern Mountain Caribou, a species that partly shares many of the challenges facing Moose and their habitat.

The Southern Mountain Caribou are a good case to use when considering the complexities of managing an ecosystem. A significant amount of the old-growth forest just outside the Wells Gray Provincial Park boundaries has been harvested. The young forests replacing the old-growth attracted Moose, and these in turn attracted wolves and other predators. Predation on the Caribou increased, while at the same time the old-growth forest tree lichens, a primary food source for Caribou when winter snow is too deep for ground lichens, diminished significantly. Forestry practices and other resource development (such as mining, logging, and oil and gas operations that resulted in new roads and forest-free corridors for transmission lines) have led to an estimated 60 per cent loss in their historic range.[151] A Caribou's diet is fairly specific, and with fewer tree lichens it went into decline.[152] Several herds have been completely lost from the area (extirpated), while others are in significant decline. Failures often happen when wildlife

MOUNTAIN CARIBOU ARE HIGHLY ENDANGERED IN SOUTHERN BC. THIS IS ONE SPECIES I'D ENCOURAGE PEOPLE TO LEARN MORE ABOUT TO BETTER UNDERSTAND THE COMPLEX REASONS AROUND THEIR DECREASING NUMBERS. A GOOD PLACE TO BEGIN IS THE FRIENDS OF WELLS GRAY PARK SOCIETY'S WEBSITE. PHOTO DAVID MOSKOWITZ – CREATIVE COMMONS.

managers are pressured to pit wildlife protection against resource extraction, with human "needs" inevitably being given superiority. The Caribou decline indicates a loss of balance within the wider ecosystem, and Moose are caught up in the mix! To understand this issue more fully, refer to the Friends of Wells Gray Park Society's website (see the Engaging and Taking Action section).

It's important to note that BC also has management successes. In the chapter on the Great Blue Heron, read about the species at risk management plan to help ensure future generations can continue to enjoy this beautiful bird. This guidebook provides numerous stories about species that are flourishing, as well as about those that are not. For instance, decisions, management practices, and recovery plans have helped to increase the numbers of Vancouver Island Marmots, Bald Eagles, and Humpback Whales. Other BC species populations are decreasing, but structured plans are coming into place to make efforts to protect these too. They include the Southern Resident Killer Whales, Western Rattlesnakes, Chinook Salmon, and several species of bats. The fact is many British Columbians are committed to protecting, conserving, and rebuilding species and their habitats. Passion for wildlife is NOT in short supply here! There are many ways to get involved in conserving BC's wildlife species, from career paths that begin at BCIT's Fish and Wildlife Management program to volunteering through your local

BC Nature naturalist club or a stewardship society such as Streamkeepers. First and foremost, inform yourself about a species in its ecosystem, such as Moose or another species you are interested in. Increasing your awareness of a species and the issues surrounding it will inevitably lead you to a greater appreciation and passion for BC's ecosystems and the species that inhabit them.

> Everything we do, we need to look through the lens of an ecosystem. Protecting just one animal without ensuring a healthy ecosystem for it to live in is just not sustainable.
>
> —**FRANK RITCEY,** naturalist, author

Engaging and Taking Action

Friends of Wells Gray Park Society is a non-profit organization set up to protect Southern Mountain Caribou habitat in the Clearwater Valley and the integrity of the wider Clearwater River watershed. Learn about the Southern Mountain Caribou herd at the BC Government's Environmental Protection

A MOOSE CALF DEVELOPS THE DISTINCTIVE FEATURES OF AN ADULT, INCLUDING LONG LEGS, BIG EARS, AND STRONG MUSCLES. PHOTO MICHAEL SCHMIDT PHOTOGRAPHY.

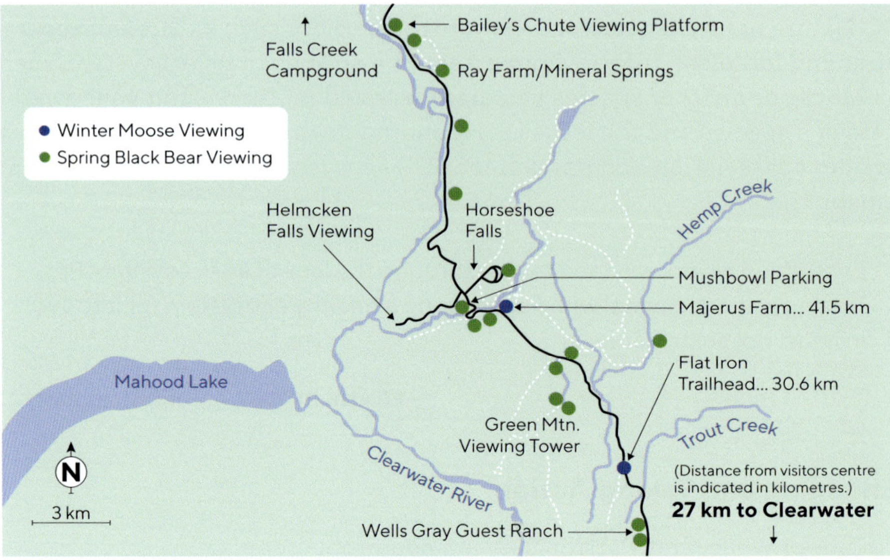

VIEWING LOCATIONS FOR BLACK BEARS (MAY–JUNE) AND MOOSE (FEBRUARY–MARCH) IN THE CLEARWATER CORRIDOR AT WELLS GRAY PROVINCIAL PARK. SOURCE: BC PARKS.

and Sustainability website, and share what you learn with others. Consider voicing your concern by writing a letter.

Check out your own local BC Nature naturalist club to see what they are involved with in your region, and get involved!

Minimizing Your Carbon Footprint

Options other than driving for getting to Wells Gray in the winter months are limited. This could be a great family trip if you pair it with staying in one of the many semi-wilderness lodges scattered throughout the central interior or in the Wells Gray corridor itself. I can't imagine children would ever forget the experience of sighting a Moose in winter or of viewing the park's stunning waterfalls (also see chapter on Black Bears in Wells Gray Park).

Directions and Accessibility

Take Highway 1 to Hope and then Highway 5 to Kamloops and on to Clearwater. The final kilometres are on the Clearwater Valley Road corridor toward Clearwater Lake. Hemp Creek and Majerus Farm are also good locations for winter Moose viewing.

Moose viewing is from a pullout and therefore fully accessible. For more details, contact the Wells Gray Park Information Centre: www.wellsgraypark.info/; info@wellsgraypark.info; (250) 674-3334.

Other Locations to View Moose

1. The Bowron Lake Canoe Circuit, reservable through BC Parks in October for the following year, is among the most stunning places to see Moose outside of winter, while on a seven-to-ten-day canoe trip.

2. Prince George near Crooked River (summer) or Grove Burn (winter). Grove Burn. The later also has the advantage of an observation tower.

3. Search for Moose along the Yellowhead Highway 16 between Prince George and Mount Robson Provincial Park, or around Cranberry Marsh near Valemount (Highway 5).

ANOTHER LARGE BC MAMMAL SPECIES IS THE ROOSEVELT ELK. THEY ARE OFTEN SEEN IN LOCATIONS NEAR SQUAMISH, ON THE SUNSHINE COAST, AND ALSO NEAR LAKE COWICHAN ON VANCOUVER ISLAND. A FAVOURITE SITE TO VIEW THEM DURING THE FALL RUT IS AT THE HOH RAIN FOREST CAMPGROUND IN OLYMPIC NATIONAL PARK IN WASHINGTON STATE. PHOTO ERWIN AND PEGGY BAUER/USFWS, FLICKR, CC PUBLIC DOMAIN MARK 1.0.

4. In Washington State, the Hoh Rain Forest at Olympic National Park provides an opportunity to view **Roosevelt Elk**, another stunningly large ungulate. In mid-September, Roosevelt Elk are in rut, and the Hoh Rain Forest Campground, surrounded by moss-covered maples and ancient cedars, is an excellent location from which to hear their "bugling."

Additional Resources

1. Differentiating Ungulates – Do you know how to tell the difference between ungulates like Moose, Elk, Caribou and Deer? There are multiple resources to see the differences online, but here's a quick summary:

    - Moose have a large dark brown body, palmate antlers, minimal tail, and a very large nose.
    - Elk have a large tan-brown body, white rump, minimal tail, and forked antlers.
    - Caribou have a brown body with a cream-coloured neck and rump, a noticeable stubby tail, and forked antlers with a lower palmate portion just above the nose.
    - Deer have the longest tail, with Mule Deer having a white tail with a black tip, and White-tailed Deer having brown on the top of the tail and white underneath. Both have antlers and large ears.

2. In both BC and Washington State there are resource fact sheets for species. Check these out through both the BC Government Nature Brochures and the Washington Department of Fish and Wildlife Fact Sheets. In them you will find local information on mammals, birds, amphibians, and reptiles.

CLIMATE CHANGE AND MOOSE

Warming temperatures due to climate change, above those to which Moose are adapted, subject Moose to heat stress in summer. BC's average temperature has increased approximately 1.4°C in the past century, with long periods of unusually hot weather in certain areas in some years (see *Indicators of Climate Change for British Columbia: 2016 Update*). Stressed Moose are more susceptible to introduced pathogens. The changing climate also affects precipitation and snow depth, and thereby food availability.

A MALE ANNA'S HUMMINGBIRD DRAWN BY THE COLOUR RED TO THIS FLOWER AS IT SEARCHES FOR NECTAR.
PHOTO SUE ELWELL

APPENDIX A

Wildlife Viewing Experiences Especially Suited for Children and Youth

Many of the wildlife viewing experiences outlined in this book are especially suitable for children, youth, and families. These are a few you may consider.

Insects – Insects offer the ability to get up close. Use chapter identification images for scavenger hunt ID. Children have a natural affinity to insects.

Winter Birds – Feeding ducks at your feet and chickadees from your hand at Reifel Bird Sanctuary is special. Watch stationary raptors such as hawks, eagles, and owls through binoculars at Boundary Bay. Hummingbirds attracted to home feeders can offer up-close and personal observation experiences as well.

Snakes – Children often have a natural affinity to reptiles and amphibians. At Nk'Mip Desert Cultural Centre, you can view snakes behind a glass enclosure, take a guided walk on snake interpretive trails, and take part in their snake program.

Intertidal Life and Tide Pooling – Tide pools provide up-close and personal observation giving children the ability to touch gently and observe behaviours of species.

Bears – Bears can be viewed safely from a car and are relatively easy to locate at the right time and place, as outlined in the chapter.

Bats – Bat-specific educational programming offered in Peachland allows you the ability to engage with bat counts, guided walks, or interpretive trails if you are visiting at a scheduled event time.

Killer Whales and Marine Mammals – The ability to see nature up close, while not guaranteed, is highly likely at the right time of year and at the right locations on the Gulf Islands. Spending full days at the beach in the summer allows the time for nature to surprise you.

Salmon or Painted Turtles – Any of BC's salmon runs are great locations for children. For certain ages, it fits directly into curriculum they learn at school. Salmon watching is multi-sensory (it smells!) and allows for up-close observation. Swan Lake Christmas Hill Nature Sanctuary is a great place to see Painted Turtles, and the boardwalk makes it easy to view them up close.

Small Mammals – Columbia Ground Squirrels are particularly entertaining to observe at the Lightning Lake day use area in Manning Provincial Park. It's an easy place to picnic at the lakeside in the summer and watch the squirrels' behaviour.

Grey Whales and Marine Mammals – A boat trip out of Tofino is sure to provide many sightings of wildlife and is likely to be an unforgettable experience for any child, particularly if whales are seen close to the boat. In addition, there are so many other family opportunities in Pacific Rim National Park, Tofino, and Ucluelet – including the aquarium!

APPENDIX B

Wildlife Viewing Experiences from Vancouver: Approximate Time Required

The wildlife viewing experiences outlined in this book require different levels of time commitment. In the following list, the trips are grouped in consideration of how much time it would take to travel from Metro Vancouver to the location (up to six hours' driving time) and then to have an unrushed, respectful, and relaxing experience with the wildlife you are hoping to see.

Day Trips:

→ Great Blue Herons in Chilliwack
→ Insects (Butterflies, Dragonflies, and Bumblebees) near Hope and Skagit Valley
→ Coyotes in Metro Vancouver
→ Winter Birds in Metro Vancouver
→ Bald Eagles in Harrison and Brackendale

Two-Day Trips (weekend):

→ Vancouver Island Marmots on Mount Washington
→ Small Mammals in Manning Provincial Park
→ Killer Whales and Marine Mammals on Saturna Island
→ Turkey Vultures at East Sooke Regional Park
→ Salmon at Goldstream Provincial Park

Three-Day Trips (long weekend):

→ Herring in the Strait of Georgia (Hornby Island)
→ Hummingbirds in Princeton
→ Grey Whales and Marine Mammals in Tofino
→ Rattlesnakes and Gopher Snakes in Osoyoos
→ Bats in Peachland
→ Intertidal Life and Tide Pooling near Port Renfrew
→ Bighorn Sheep at Lac du Bois, near Kamloops

Four-Day Trips (extra-long weekend):

→ Bears in Wells Gray Provincial Park
→ Meadowlark Nature Festival in the Okanagan
→ Moose in Wells Gray Provincial Park

APPENDIX C

Evolutionary History of Selected Species Covered in *Wildlife Weekends*

New Zealand was the last landmass in the world to be settled by humans when the Maori sailed there about 700 years ago, followed by Europeans about 400 years ago. In 2019, I visited New Zealand, and as I learned about New Zealand's flora and fauna, and the changes they have endured since humans arrived, it became apparent that their Department of Conservation projects and actions were being taken to protect species that had been evolving on that geographically isolated island for not just thousands but millions of years!

BC may not have the same isolated geography that New Zealand does, but I knew I wanted to understand the depth of evolutionary history for some of the species I've profiled, to put into context the scale of our human enterprise in relation to them. Species described in this book are the unique result of millions of years of genetic adaptation and evolution, so when we observe an animal in its habitat, we are looking at an organism that has, just like us, evolved over time to survive, and in fact thrive, in its particular niche. And they have flourished right through until this juncture in geologic time we call the **Holocene epoch**. This time period corresponds to approximately the last 11,000 years, but have you heard yet about the **Anthropocene**? There is strong evidence that suggests human actions have shifted the planet dangerously out of the Holocene and into a new epoch many are calling the Anthropocene, in which humans have become the main influencers of Earth's climate and ecosystems. (There is no firm agreement yet as to when exactly the Anthropocene began, although suggestions include when humans moved to agriculture thousands of years ago; the Industrial Revolution, which started in the late 18th century; the age of nuclear tests that spread radioactive elements around the planet, beginning in the 1940s; and the **Great Acceleration**, the dramatic and continuous surge from the mid-1950s onward in economic growth across a large range of measures of human activity, continuing to this day.) Many believe the blame belongs to pro-capitalism policies in place in many parts of the world, which have prioritized resource extraction and unfettered development, endangering many species of wildlife and pushing the planet's climate and ecosystems to the limit of survivability. Today, it is imperative that we humans change our ways and learn to respect the **Planetary Boundaries** (an important concept I think everyone should become familiar with), and repair our relationship to the natural world, making biodiversity a priority. If we don't, we certainly won't flourish, and, at worst, society as we know it won't survive.

BC has many species that are closer to extinction or **extirpation** (removal or extinction from a particular geographic area, but continuing to exist elsewhere), than you might think. *And extinction is forever.* Some of the species we have brought to the brink of extinction or extirpation are included in this book: the Vancouver Island Marmot, the Southern Resident Killer Whales, and the Yellow-breasted Chat. Some are not included here, either because they don't exist anymore or because there are just too few of them left to call a population: the Northern Spotted Owl, Southern Mountain Caribou herds, and soon, possibly the Taylor's Checkerspot Butterfly. My hope is that the table below will help the reader understand how long these species have been around, and how little time it has taken us to drastically reduce their populations.

SPECIES	Evidence showing the species has existed since at least...
Grey Whale	Flourished since the late Pleistocene (the geological epoch that lasted from about 2,580,000 to 11,700 years ago, spanning the world's most recent period of repeated glaciations). Grey Whales descended from a common filter-feeding whale ancestor around **30 million years ago**.
Hummingbird	Here in the New World for at least 10 million years and moving into North America during the rise of the land bridge in Panama three million years ago. Existing hummingbirds descended from a common ancestor around **22 million years ago**.

Sandhill Crane	These have one of the longest fossil histories of any bird. The oldest unequivocal Sandhill Crane fossil is 2.5 million years old, with a close ancestor perhaps **10 million years ago**.
Raven	Common Ravens settled in California at least two million years ago. They became separated from their closest relatives in Europe and Asia during a previous ice age, also around **two million years ago**.
Black Bear	The American Black Bear split from the Asian Black Bear just over **four million years ago**. Though other bear species existed in North America, it alone survived climate and vegetative changes through the last ice age, until the later migration of Grizzly Bears.
Coyote	Coyotes diverged from their common ancestor in the middle Pleistocene, about **one million years ago**. All North American wolves have a significant amount of Coyote ancestry, and all Coyotes have some degree of wolf ancestry.
Moose	This early Pleistocene epoch animal is a relative latecomer to North America, crossing the Bering land bridge from Asia. It diverged from its common ancestor around **two million years ago**.
Bighorn Sheep	Bighorn Sheep originally entered North America from Siberia, crossing over the Bering land bridge from Asia in the Pleistocene epoch. They diverged from their closest Asian relative about **600,000 years ago**. At one time, millions likely roamed throughout most of North America, ranging as far south as central Mexico.
Salmon	Scientists use both the fossil record and analysis of mitochondrial DNA to see that the last divergence of Pacific from Atlantic salmon occurred **10 to 20 million years ago**, long before the glaciers. (Quaternary glaciation beginning 2.58 million years ago began their cycle of advance and retreat.)
Humans	Homo sapiens evolved from our common ancestor around just **300,000 years ago**.

And Others**:** **Bats** are from the Eocene **(65–39 million years ago [mya])**, **Marmots** the Miocene **(23–5.3 mya)**, **Orcas** the Pliocene **(5.3–2.5 mya)**, and **Turtles** are from the middle Jurassic **(201–145 mya!)**.

THE AMERICAN PIKA IS A MEMBER OF THE RABBIT FAMILY AND IS OFTEN FOUND AMONGST LARGE ROCK OUTCROPS AT HIGH ELEVATIONS. LOOK FOR THEIR BIG EARS ON SMALL BODIES.
PHOTO CHRISTINE GAIO.

APPENDIX D

Wildlife-Related United Nations and Canadian Recognized Days

What day is it today and how many more days are there until the next Internationally or nationally recognized day that relates to wildlife? Often there are in-person or online events associated with these days. Both Canada and the United Nations have recognized special days, and many of them revolve around wildlife and nature. In addition, I have added a few extra days that revolve around wildlife viewers themselves.

2 February	World Wetlands Day
3 March	World Wildlife Day
21 March	International Day of Forests
22 March	World Water Day
23 March	World Meteorological Day
10 April	Canadian National Wildlife Week
22 April	International Mother Earth Day
First Sunday in May	International Dawn Chorus Day
9–10 May	World Migratory Bird Day
20 May	World Bee Day
22 May	International Day for Biological Diversity

5 June	World Environment Day & Canadian Environment Week
8 June	World Oceans Day
21 June	National Indigenous Peoples Day
16 July	World Snake Day
18 July	World Listening Day
30 September	National Day for Truth and Reconciliation
3 October	World Habitat Day
5 December	World Soil Day
11 December	International Mountain Day

Related to Wildlife Viewers

15 May	International Day of Families
15 July	World Youth Skills Day
12 August	International Youth Day
20 November	Universal Children's Day
3 December	International Day of Persons with Disabilities
19 April	National Volunteer Week
5 September	International Day of Charity

REFERENCES

List of Referenced Resources & Organizations with Hyperlinks

Introduction

3billionbirds.org - https://www.3billionbirds.org/
Three C's - https://www.orcbc.ca/responsible-recreation
Leave No Trace - https://www.leavenotrace.ca/home
BC Parks Responsible Recreation Guidelines - https://bcparks.ca/visiting/responsible-recreation/
1000clearcuts.ca - https://1000clearcuts.ca/
Caribou Extinction - https://thenarwhal.ca/a-sad-day-two-more-b-c-mountain-caribou-herds-now-locally-extinct/
Sixth Global Extinction - https://en.wikipedia.org/wiki/The_Sixth_Extinction:_An_Unnatural_History
Nature Editorial Quote - https://www.nature.com/articles/541260a

Pacific Herring in the Strait of Georgia & the Hornby Island HerringFest

Conservancy Hornby Island - https://www.conservancyhornbyisland.org/
HerringFest - https://www.conservancyhornbyisland.org/herringfest2024
Hakai Herring School - http://www.pacificherring.org/timeline
It's Complex! - https://www.youtube.com/watch?v=BwjBYAOd2IM
DFO Fishery Notices - https://www-ops2.pac.dfo-mpo.gc.ca/fns-sap/index-eng.cfm

13 Moons of the W̱SÁNEĆ - https://racerocks.ca/the-13-moons-of-the-wsanec/
Project Watershed - https://projectwatershed.ca/2020/03/09/forage-fish/
SeaChange Society - https://seachangesociety.com/get-involved/
Tourism Vancouver Island - https://forvi.ca/
Herring Roe Fishery, Strait of Georgia 2020 - https://www.youtube.com/watch?v=6EzO5wugg4Y
Shifting Baselines - https://www.ted.com/talks/daniel_pauly_the_ocean_s_shifting_baseline
Raincoast Conservation Project - https://www.raincoast.org/projects/marine-birds/pacific_herring/

Grey Whales & Marine Mammals in Clayoquot Sound & Pacific Rim National Park, Vancouver Island's West Coast

Pacific Rim Whale Festival - https://www.pacificrimwhalefestival.com/
Ocean Wise Whales - https://ocean.org/whales/
Ocean Wise Whales Printable Guide - https://ocean.org/action/identification-guides/
UNESCO Clayoquot Biosphere Reserve - https://clayoquotbiosphere.org/
Nuu-chah-nulth Language Learning - https://www.firstvoices.com/nuu-chah-nulth-barkley/
West Coast N.E.S.T. - https://www.westcoastnest.org/
Clayoquot Biosphere Trust - http://clayoquotbiosphere.org/
Pacific Rim National Park - https://www.pc.gc.ca/en/pn-np/bc/pacificrim
Wild Pacific Interpretive Trail - http://wildpacifictrail.com/home.html
Ucluelet Aquarium - https://uclueletaquarium.org/
Be a Better Beach Goer - https://uclueletaquarium.org/wp-content/uploads/be-a-better-beach-goer.pdf
Tofino Shorebird Festival - https://raincoasteducation.org/what-we-do/events/tofino-shorebird-festival/
Surfrider Foundation - https://vancouverisland.surfrider.org/
Marine Debris Tracker Mobile App - https://blog.marinedebris.noaa.gov/marine-debris-tracker-fight-marine-debris-your-phone
The Ocean Cleanup - https://theoceancleanup.com/
Strawberry Isle Marine Research Society - https://www.simrstofino.org/our-team
Happywhale Whale Sightings - https://www.happywhale.com/home
DFO Marine Mammal Incident Reporting Hotline - https://dfo-mpo.gc.ca/species-especes/mammals-mammiferes/report-rapport/page01-eng.html
International Phonetic Alphabet - https://www.internationalphoneticalphabet.org/ipa-sounds/ipa-chart-with-sounds/
Unusual Mortality Event - https://www.fisheries.noaa.gov/national/marine-life-distress/2019-2021-gray-whale-unusual-mortality-event-along-west-coast-and

Pacific Great Blue Herons at the Great Blue Heron Nature Reserve in Chilliwack

Chilliwack Great Blue Heron Nature Reserve - http://chilliwackblueheron.com/

Allaboutbirds.org - https://www.allaboutbirds.org/guide/Great_Blue_Heron/overview

COSEWIC (Committee on the Status of Endangered Wildlife in Canada) - http://www.cosewic.ca/index.php/en-ca/

Species at Risk Legislation - https://www2.gov.bc.ca/gov/content/environment/plants-animals-ecosystems/species-ecosystems-at-risk/legislation

Cowichan Estuary Nature Centre Heron Cam - https://www.cowichanestuary.ca/heron-camera-2020/

Stanley Park Ecology Society's Heron Cam - https://vancouver.ca/parks-recreation-culture/heron-cam.aspx

Stanley Park Ecology Society's Adopt a Heron Nest - https://stanleyparkecology.ca/support-us/adopt/

Hummingbirds in Princeton along the Kettle Valley Railway Trail

State of Canada's Birds 2019 Report - http://nabci.net/resources/state-of-canadas-birds-2019/

Rocky Point Bird Observatory's Hummingbird Project - http://rpbo.org/hummingbirds/

Rocky Point Bird Observatory - https://rpbo.org/frequently-asked-questions/

Birds Canada - https://www.birdscanada.org/

Otter Lake Provincial Park - https://bcparks.ca/explore/parkpgs/otter_lk/

Project Feeder Watch - https://www.birdscanada.org/you-can-help/project-feederwatch/

Great Backyard Bird Count - https://www.birdscanada.org/bird-science/great-backyard-bird-count/

Discover Princeton - https://discoverprinceton.ca/

Kettle Valley Railway Trail - https://bcrailtrails.com/

The Trans Canada Trail - https://tctrail.ca/

Vancouver Paleontology Society - http://www.vcn.bc.ca/vanps/

Bromley Rock Provincial Park - http://bcparks.ca/explore/parkpgs/bromley_rock/

Hedley's historic Mascot Mine - https://www.youtube.com/watch?v=C7lo3voLHbc

IUCN Red List (International Union for Conservation of Nature) - https://www.iucnredlist.org/species/22688296/131750335

Black Bears in the Clearwater Corridor at Wells Gray Provincial Park

Grouse Mountain Grizzly Bears Infrared Webcam - https://www.grousemountain.com/webcam-winter/bear-den-cam

Clearwater Corridor Map - https://sci-bc.ca/wp-content/uploads/2024/05/Wells-Gray.pdf

WildSafeBC - https://wildsafebc.com/

The Salmon Forest - https://web.uvic.ca/~reimlab/salmonforest.html

Bear Smart Community Program - https://www2.gov.bc.ca/gov/content/environment/plants-animals-ecosystems/wildlife/human-wildlife-conflict/staying-safe-around-wildlife/bears/bear-smart

W.A.R.P. - Wildlife Alert Reporting System - https://wildsafebc.com/programs/what-is-warp/

BC Wildlife Federation Conservation Mobile App - https://bcwf.bc.ca/initiatives/conservation-app/

BC Parks iNaturalist Project - https://bcparksfoundation.ca/projects/wildlife-forever/bc-parks-inaturalist-project/

BC Parks Foundation - https://bcparksfoundation.ca/blog/1-million-observations-of-wildlife-in-b.c/

Thornton Creek Hatchery Facebook Page - https://www.facebook.com/ThorntonCreekHatchery

AdventureSmart - https://www.adventuresmart.ca/index.php

The Meadowlark Nature Festival in the South Okanagan & Similkameen

Okanagan Similkameen Conservation Alliance - https://osca.org/

Seven Simple Actions to Help Birds - https://www.birds.cornell.edu/home/seven-simple-actions-to-help-birds/

One Square Inch of Silence - https://onesquareinch.org/about/

Vaseux Lake Bird Observatory - https://osca.org/vaseux-lake-bird-observatory/

Okanagan Similkameen Stewardship - https://www.osstewardship.ca/projects

Vancouver Bird Celebration - https://vancouverbirdcelebration.ca/

Tofino Shorebird Festival - https://raincoasteducation.org/what-we-do/events/tofino-shorebird-festival/

Brant Wildlife Festival - https://brantfestival.bc.ca/

Wings Over Water - https://www.wingsoverwaterbirdingfestival.com/

Wings Over the Rockies - http://www.wingsovertherockies.org/

30 by 30 Target - https://naturecanada.ca/defend-nature/how-you-help-us-take-action/nature-based-climate-solutions/toolkit/protection/

Wildlife Viewing in the South Okanagan and Similkameen Brochure - http://www.env.gov.bc.ca/fw/wildlife/viewing/docs/wvokanagan.pdf

The Voice of the Natural World - https://www.ted.com/talks/bernie_krause_the_voice_of_the_natural_world
Earth Day Wetland Project - http://www.coopradio.org/events/wild-creatures-take-over-airwaves-earth-day-april-22
BC Government Sensitive Ecosystem Inventory - https://www2.gov.bc.ca/gov/content/environment/plants-animals-ecosystems/ecosystems/search-ecosystem-info

Vancouver Island Marmots at Mount Washington, Strathcona Provincial Park

IUCN Critically Endangered Red List - https://www.iucnredlist.org/species/12828/22259184
Marmot Recovery Foundation - https://marmots.org/
Sustainable Development Goal #17 Partnerships - https://sdgs.un.org/goals/goal17
Marmot Recovery Plan - https://marmots.org/recovery-efforts/recovery-strategy/
Strathcona Provincial Park - https://bcparks.ca/explore/parkpgs/strath/
Adopt-a-Marmot - https://marmots.org/how-you-can-help/adopt-a-marmot/
Mount Washington Alpine Resort - https://www.mountwashington.ca/
Mount Arrowsmith Regional Park - https://www.rdn.bc.ca/mount-arrowsmith-massif-regional-park
Haley Lake Ecological Reserve - https://bcparks.ca/eco_reserve/haley_er.html
Courtenay Museum and Paleontology Centre and Fossil Tours - https://www.courtenaymuseum.ca/
The Sustainable Development Goals (SDGs) Progress Report - https://dashboards.sdgindex.org/
New Marmot Colony Found - https://marmots.org/category/news-from-the-field/page/2/

Insects: Butterflies, Dragonflies, & Bumblebees in Skagit Valley Provincial Park & Cheam Lake Wetlands Regional Park

The Butterflyway Project - https://davidsuzuki.org/take-action/act-locally/butterflyway/
City of Vancouver Green Streets Program - https://vancouver.ca/home-property-development/green-streets-program.aspx
Cheam Lake Wetlands Regional Park - https://www.fvrd.ca/EN/main/parks-recreation/parks-trails/cheam-lake-wetlands.html
Skagit Valley Provincial Park - https://bcparks.ca/explore/parkpgs/skagit/
E-Fauna BC - https://ibis.geog.ubc.ca/biodiversity/efauna/index.shtml
E-Flora BC - https://ibis.geog.ubc.ca/biodiversity/eflora/
Biodiversity of BC - https://ibis.geog.ubc.ca/biodiversity/

Royal BC Museum Dragonflies of BC - https://publications.royalbcmuseum.bc.ca/product/introducing-the-dragonflies-of-british-columbia-and-the-yukon/
Bee City Canada - https://beecitycanada.org/bee-cities/
Bumblebeewatch - https://www.bumblebeewatch.org/
iNaturalist.ca - https://inaturalist.ca/
Spencer Entomological Collection - Beaty Biodiversity Museum - https://beatymuseum.ubc.ca/research-2/collections/spencer-entomological-collection/
BC Wild Heritage Plants - http://www.bcwildheritage.com/
West Coast Seeds - https://www.westcoastseeds.com/
Monarchbutterflies.ca - https://www.monarchbutterflies.ca/
Bumble Bees of the Western United States - https://xerces.org/publications/identification-and-monitoring-guides/bumble-bees-of-western-united-states
We're in a biodiversity crisis. What we plant and how we alter landscapes matter - https://www.natureconservancy.ca/en/blog/archive/were-in-a-biodiversity-crisis.html

Western Rattlesnakes & Great Basin Gopher Snakes at the Nk'Mip Desert Cultural Centre, Osoyoos, South Okanagan

Nk'Mip Desert Cultural Centre - https://nkmipdesert.com/
Osoyoos Indian Band - https://okib.ca/
Seek App - https://www.inaturalist.org/pages/seek_app
Adopt-a-Snake - https://www.nkmipdesert.com/protecting-desert-dwellers
Okanagan Stewardship Society - https://www.osstewardship.ca/get-involved
Okanagan Similkameen Conservation Alliance - https://osca.org/wp-content/uploads/2017/08/sos_snakes.pdf
Help Stop the Spread of Invasive Species - Prevention Tips - https://bcinvasives.ca/wp-content/uploads/2021/01/Prevention_Tips_FINAL_18_04_2019.pdf
The Desert Centre - https://www.desert.org/
White Lake Grasslands Protected Area - https://bcparks.ca/explore/parkpgs/wht_lk_grass/
Kentucky-Alleyne Provincial Park - https://bcparks.ca/explore/parkpgs/kentucky_alleyne/
BCreptiles.ca - http://www.bcreptiles.ca/
Biodiversity Research Centre at UBC - https://biodiversity.ubc.ca/

Small Mammals (Squirrels, Marmots, & American Pikas) at E.C. Manning Provincial Park

Manning Provincial Park - https://bcparks.ca/explore/parkpgs/ecmanning/
BC Park iNaturalist - https://www.inaturalist.org/projects/e-c-manning-provincial-park
Hope Mountain Centre - http://hopemountain.org/
Hudson's Bay Company Heritage Trail - https://hopemountain.org/trails/hbc-heritage-trail/
Manning Park Resort - https://www.manningpark.com/
Coquihalla Canyon Provincial Park - https://bcparks.ca/explore/parkpgs/coquihalla_cyn/
Hinterland Who's Who - https://www.hww.ca/en/index.html
BC Parks Wildlife Viewing Brochure - https://nrs.objectstore.gov.bc.ca/kuwyyf/wildviewing_part2_b5d8ee5a8a.pdf
Nature Conservancy of Canada - https://www.natureconservancy.ca/en/
Summer Tracking for Beginners - https://www.youtube.com/watch?v=YYz47DaY1os

Yuma & Little Brown Bats in Peachland on Okanagan Lake

Okanagan Similkameen Conservation Alliance - https://osca.org/living-with-wildlife/
Bat Friendly Community - https://www.bcbats.ca/index.php/get-involved/bat-friendly-communities
Bat House Interpretive Trail - https://beepspeachland.com/programs/
Neonic-free Plants - https://www.wildernesscommittee.org/news/where-go-plants-season
Wildlife Conservation Society Canada - https://wcsbats.ca/probiotic
BCBat.ca Bat Counts - https://www.bcbats.ca/attachments/count-instructions-2020.pdf
BCbats.ca Bat House - https://bcbats.ca/bat-boxes/
BC Nature Local Naturalist Organizations - https://bcnature.org/bc-nature-clubs/
Peachland Tourism Visitors Centre - https://visitpeachland.com/
International Bat Week - https://batweek.org/
COSEWIC Recovery Strategies - https://www.canada.ca/en/environment-climate-change/services/species-risk-public-registry/recovery-strategies/little-brown-myotis-2018.html#toc12
Kelowna Museum Bat Exhibit - https://www.kelownamuseums.ca/exhibits/bats/
US Fish and Wildlife Bats Brochure - https://woodlandfishandwildlife.com/wp-content/uploads/2019/07/BatsAndSmallWoodland.pdf
Canadian Wildlife Federation's Bat House Survey - https://cwf-fcf.org/en/explore/bats/bat-survey-1.html

Urban Coyotes & Other Wildlife in Metro Vancouver

Stanley Park Ecology Society's Co-Existing with Coyotes Program - http://stanleyparkecology.ca/conservation/co-existing-with-coyotes/coyote-sighting/
BCSPCA - https://spca.bc.ca/ways-to-help/take-action/urban-wildlife/
WildSafeBC - https://wildsafebc.com/species/coyote/
Wildlife Alert Reporting Program - https://wildsafebc.com/programs/what-is-warp/
Predator Conflicts & Statistics - https://www2.gov.bc.ca/gov/content/environment/plants-animals-ecosystems/wildlife/human-wildlife-conflict
Wildlife Rescue Association of BC - https://www.wildliferescue.ca/
Streamkeepers Groups - http://www.streamkeepers.info/
MyImpactPage at Metro Vancouver Parks - https://app.betterimpact.com/PublicOrganization/46f4140f-db5f-42be-ac8a-b0e65c38c382/1
Beaty Biodiversity Museum - https://beatymuseum.ubc.ca/
Stanley Park Nature House - https://stanleyparkecology.ca/education/nature-house/
Coyote Watch Canada - https://www.coyotewatchcanada.com/site/home
BC Conservation Officer Service -https://www2.gov.bc.ca/gov/content/environment/natural-resource-stewardship/natural-resource-law-enforcement/conservation-officer-service
Ministry of Environment and Climate Change - https://www2.gov.bc.ca/gov/content/environment/plants-animals-ecosystems/wildlife/human-wildlife-conflict/staying-safe-around-wildlife
Climate 2050: A Climate Strategy for the Metro Vancouver Region - https://metrovancouver.org/services/air-quality-climate-action/Documents/climate-2050-strategic-framework-2018.pdf
Active Vancouver: A Year-Round Guide to Outdoor Recreation in the City's Natural Environments - https://rmbooks.com/book/active-vancouver/
Urban Coyote Research Project - https://urbancoyoteresearch.com/

Intertidal Life & Tide Pools at Botanical Beach, Juan de Fuca Provincial Park

Tidal Prediction Charts for Port Renfrew - tides.gc.ca - https://tides.gc.ca/en/stations/08525
Bamfield Marine Sciences Centre - https://www.bamfieldmsc.com/bmsc-overview
West Coast N.E.S.T. - https://www.westcoastnest.org/
Clayoquot Biosphere Trust - http://clayoquotbiosphere.org/
Ucluelet Aquarium - https://uclueletaquarium.org/
Ocean Wise - https://ocean.org/about/
Ocean Bridge - https://bridge.ocean.org/
High Level Panel for a Sustainable Ocean Economy - https://oceanpanel.org/

Sustainable Development Goal #14 Life Below Water - https://www.undp.org/sustainable-development-goals#below-water
Surfrider Foundation - https://www.surfrider.org/
Metro Vancouver Switch to Cold - https://metrovancouver.org/microfibres/
Seastarwasting.org - https://marine.ucsc.edu/data-products/sea-star-wasting/index.html
Howe Sound Biosphere Regional Initiative - https://www.howesoundbri.org/
OceanWatch - https://oceanwatch.ca/howesound/wp-content/uploads/sites/2/2016/12/OceanWatch-HoweSoundReport-SeaStars.pdf
Seafood Watch - https://www.seafoodwatch.org/globalassets/sfw/pdf/guides/seafood-watch-west-coast-guide.pdf
West Coast Trail Express - https://trailbus.com/
Sooke – Port Renfrew Tourism - https://sooke-portrenfrew.com/
Juan de Fuca Marine Trail - http://vancouverisland.com/things-to-do-and-see/parks-and-trails/trails/juan-de-fuca-marine-trail/
Juan de Fuca Provincial Park - https://bcparks.ca/explore/parkpgs/juan_de_fuca/trailhd.html
The Global Hub for Real-Time SDG Data - https://sdgstoday.org/

Killer Whales (*Orcinus orca*) & Marine Mammals: Shore-based Viewing from East Point Lighthouse on Saturna Island

W̱SÁNEĆ people of the Saanich Peninsula - https://wsanec.com/
Narvaez Bay at tides.gc.ca - https://tides.gc.ca/eng/station?sid=7345
WildWhales.org Printable Guide - https://oceanorg.blob.core.windows.net/oceanorg/2024/01/OWSN_IDGuides_2024.pdf
Killer Whales - https://www.whaleresearch.com/appid?utm_source=so&cid=9ab1afc8-1178-48ce-8ab8-a329f460db85
Centre for Whale Research - https://www.whaleresearch.com/?cid=9ab1afc8
Current Number of SRKWs - https://www.orcanetwork.org/births-and-deaths
Saturna Island Marine Research and Education Society (SIMRES) - https://simres.ca/
J,K & L Pods - https://www.whaleresearch.com/orca-population
Fog Alarm Building - http://saturnaheritage.ca/
Narvaez Bay Campsite - https://www.pc.gc.ca/en/pn-np/bc/gulf/activ/camping
Marinetraffic.com - https://www.marinetraffic.com/en/ais/home/centerx:-123.1/centery:48.5/zoom:10
Whale Watching Regulations - 400 metres - https://www.bewhalewise.org/regulations-guidelines/
Sustainable Development Goals - https://sdgs.un.org/goals
SDG Target 4.7 - https://sdgs.un.org/goals/goal4

BC Cetacean Sightings Network - https://ocean.org/overfishing/whales/ocean-wise-sightings-network/
WhaleReport Mobile App - https://play.google.com/store/apps/details?id=ca.org.vanaqua.vawhalereport&hl=en_CA
SeaChange Society Plant & Restore Eelgrass - https://seachangesociety.com/
SeaChange opportunities to volunteer - https://seachangesociety.com/get-involved/
Mayne Island Conservancy - https://mayneconservancy.ca/kelp-bed-monitoring/
Marine Life Sanctuaries Society - https://mlssbc.com/
SIMRES Sea Talks - https://simres.ca/seatalks/
BC Coast Hydrophone Network - https://www.bcwhales.org/bchydrophonenetwork
Orca-live.net - http://www.orca-live.net/community/index.html
Real-time forecasting system for SRKWs - https://www.sfu.ca/~mrandon/SRKW_project.html
Saturna Island Tourism Association - https://www.saturnatourism.com/
Lime Kiln Point, San Juan Island - https://www.parks.wa.gov/540/Lime-Kiln-Point
Friends of Lime Kiln Society - http://www.folkssji.org/
The Whale Trail - https://thewhaletrail.org/
Happywhale - https://happywhale.com/home
Georgia Strait Alliance - https://georgiastrait.org/orca-action/
Cetus Society's Straitwatch - https://www.cetussociety.org/straitwatch
Raincoast Conservation Foundation - https://www.raincoast.org/2021/10/whale-watching-operators-recreational-fishers-routinely-violating-protection-zone-buffers-srkw/
Port of Vancouver's Terminal 2 Expansion - https://www.raincoast.org/2020/07/terminal-2-backgrounder-southern-resident-killer-whales/
ECHO Program (Enhancing Cetacean Habitat and Observation) - https://www.portvancouver.com/environmental-protection-at-the-port-of-vancouver/maintaining-healthy-ecosystems-throughout-our-jurisdiction/echo-program/
Hakai Magazine - https://hakaimagazine.com/features/trail-whales/
SeaDoc Society's Salish Sea Wild Videos - https://www.seadocsociety.org/
Marine Education and Research Society - https://mersociety.org/

Turkey Vultures & Raptors at East Sooke Regional Park, Vancouver Island

East Sooke Regional Park - https://www.crd.bc.ca/parks-recreation-culture/parks-trails/find-park-trail/east-sooke
Merlin ID Bird Mobile App - https://merlin.allaboutbirds.org/
Allaboutbirds.org - https://www.allaboutbirds.org/guide/browse/taxonomy/Accipitridae

Cornell Lab Bird Academy - https://academy.allaboutbirds.org/
Audubon's Identify Raptors in Flight - https://www.audubon.org/magazine/fall-2021/look-introduction-identifying-raptors-flight
Victoria Natural History Society - http://www.vicnhs.bc.ca/
Rocky Point Bird Observatory - https://rpbo.org/
Audubon's Survival by Degrees: 389 Species on the Brink - https://www.audubon.org/climate/survivalbydegrees
Animal Tracker Mobile App - https://www.icarus.mpg.de/29143/animal-tracker-app
Cornell Lab's Three Billion Birds - https://www.birds.cornell.edu/home/bring-birds-back/
Birds Canada - https://www.birdscanada.org/
Lochside Trail & Galloping Goose Trail - http://www.gallopinggoosetrail.com/pdf/gallop_goose_info.pdf
Sooke – Port Renfrew Tourism - https://sooke-portrenfrew.com/
Sooke Potholes Provincial Park - https://bcparks.ca/explore/parkpgs/sooke_potholes/
Roche Cove and Matheson Regional Parks - https://www.crd.bc.ca/parks-recreation-culture/parks-trails/find-park-trail/roche-cove
Whiffen Spit Park - https://www.victoriatrails.com/trails/whiffin-spit/
Tibetan Sky Burials - https://en.wikipedia.org/wiki/Sky_burial

Chum, Coho ,& Chinook Salmon in Goldstream Provincial Park, Vancouver Island

Goldstream Provincial Park - https://bcparks.ca/explore/parkpgs/goldstream/
Chum Salmon Spawn! - https://www.youtube.com/watch?v=lQO9KdG-3Ko
Tire Rubber Shed - https://www.hakaimagazine.com/features/when-rubber-hits-the-road-and-washes-away/
Goldstream Park Nature House (Freeman King Visitor Centre) - https://www.goldstreampark.com/freeman_king_visitor_center.htm
Streamkeepers (Burnaby) - https://burnabylakepark.ca/burnaby-streamkeeper-groups/
Watershed Watch Salmon Society - https://watershedwatch.ca/volunteer/
Pacific Salmon Foundation - https://www.psf.ca/
Squamish River Watershed Society - https://www.squamishwatershed.com/
Fraser Valley Conservancy - https://fraservalleyconservancy.ca/volunteer-opportunities/
The Reptiles of British Columbia - https://bcreptilesandamphibians.ca/western-painted-turtle/
Wildlife at Risk - http://www.env.gov.bc.ca/wld/documents/pturtle.pdf
British Columbia Marsh Monitoring Program - https://birdscanada.org/birdmon/bcmmp/resources.jsp?

Lochside and Galloping Goose Regional Trails - http://www.gallopinggoosetrail.com/pdf/lochside_trail_guide.pdf
Coquitlam's Salmon Come Home Map - https://www.coquitlam.ca/DocumentCenter/View/10912/2023-Salmon-Come-Home-Scavenger-Hunt-
Southern BC's Salmon Spawning Sites & Times - https://watershedwatch.ca/stories/where-to-see-salmon-spawning/
Capilano River Hatchery - https://www.pac.dfo-mpo.gc.ca/sep-pmvs/hatcheries-ecloseries/capilano-eng.html
Weaver Creek Spawning Channel - https://www.pac.dfo-mpo.gc.ca/sep-pmvs/hatcheries-ecloseries/spawning-channel-frayeres-artif/weaver-eng.html
Swan Lake - Christmas Hill Nature Sanctuary - https://www.swanlake.bc.ca/
Red-listed Species - https://www2.gov.bc.ca/gov/content/environment/plants-animals-ecosystems/conservation-data-centre/explore-cdc-data/red-blue-yellow-lists
Pacific Salmon Explorer - https://salmonexplorer.ca/#!/
Raincoast Conservation Foundation's Habitat Restoration in the Fraser River Estuary - https://www.raincoast.org/fraser-river-estuary-project/
Pacific Salmon and Their Circle of Life (Bob Turner video) - https://www.youtube.com/watch?v=PHKsXQrl41c
Seymour River Fish Hatchery & Education Centre - http://seymoursalmon.com/

California Bighorn Sheep at Lac du Bois Grasslands Protected Area, Kamloops

Lac du Bois Grasslands Protected Area - https://bcparks.ca/explore/parkpgs/lacdubois_grass/
BC Sheep Separation Program - https://hctf.ca/bc-s-wild-domestic-sheep-separation-program/
Guidelines for Ethical Photography - https://www.raincoast.org/photography/
Johnson Lake Instagram Effect - https://www.cbc.ca/news/canada/british-columbia/johnson-lake-destruction-overcrowding-blamed-on-social-media-1.3151542
Large Carnivore Hunting and the Social Licence to Hunt - https://www.raincoast.org/2021/02/research-the-eroding-social-license-to-hunt-carnivores/
BC Provincial RAPP Line - https://www2.gov.bc.ca/gov/content/environment/natural-resource-stewardship/natural-resource-law-enforcement/conservation-officer-service/cos-rapp
BC Nature Photography Contest - https://bcnature.org/photo-contest/
BC Wildlife Federation - https://bcwf.bc.ca/
Raincoast Conservation Foundation - https://www.raincoast.org/photography/
Hinterland Who's Who - https://www.hww.ca/en/wildlife/mammals/mountain-sheep.html

Winter Birds in Metro Vancouver

Westham Island's Reifel Bird Sanctuary - https://www.reifelbirdsanctuary.com/

Cornell Lab's Merlin Birding App - https://merlin.allaboutbirds.org/

Nature Vancouver's Seasonal Checklist of Greater Vancouver's Birds - https://www.naturevancouver.ca/wp-content/uploads/2017/04/Birds-of-Greater-Vancouver-Checklist.pdf

BC Bird Trail - https://bcbirdtrail.ca

BC Nature's Local Naturalist Clubs & Organizations - https://bcnature.org/bc-nature-clubs/

BC Nature Guide - https://bcnature.org/bc-nature-clubs/

Delta Naturalists Series of Nature Brochures - https://dncb.wordpress.com/dns-brochures/

Breeding Bird Survey - https://www.birdscanada.org/bird-science/breeding-bird-survey

British Columbia Coastal Waterbird Survey - https://www.birdscanada.org/bird-science/british-columbia-coastal-waterbird-survey

UBC's Biodiversity Research Centre - https://biodiversity.ubc.ca/

Flying by the Fat of the Sea - https://www.hakaimagazine.com/features/flying-by-the-fat-of-the-sea/

BC Breeding Bird Atlas - https://www.birdatlas.bc.ca/

Cornell Lab's Birdcast - https://birdcast.info/

Vancouver Bird Celebration - https://vancouverbirdcelebration.ca/

World Migratory Bird Day - https://www.migratorybirdday.org/

Reifel Bird Sanctuary's Rules for Feeding Birds - https://www.reifelbirdsanctuary.com/feeding.html

Seymour River Fish Hatchery and Education Centre - https://seymoursalmon.com/

Minnekhada Regional Park - https://metrovancouver.org/services/regional-parks/park/minnekhada-regional-park

Iona Beach Regional Park - https://metrovancouver.org/services/regional-parks/park/iona-beach-regional-park

City of Vancouver Parks Guide - https://vancouver.ca/files/cov/vancouver-parks-map-and-guide.pdf

BC Rare Bird Alert - https://bcbirdalert.blogspot.com/

Maplewood Flats Conservation Area - https://wildbirdtrust.org/visit-maplewood-flats/

A Nature Guide to Boundary Bay - https://www.natureguidesbc.com/Boundary_Bay.html

Delta Naturalists Nature in Delta Brochures - https://dncb.wordpress.com/dns-brochures/

Key Biodiversity Areas - http://www.keybiodiversityareas.org/about-kbas

Hawk Migration Association of North America's HawkCount - https://hawkcount.org/sitesel.php

Seattle Audubon's Guide to the Birds of Washington State BirdWeb - http://birdweb.org/birdweb/
Seven Simple Actions to Help Birds - https://www.birds.cornell.edu/home/seven-simple-actions-to-help-birds/
Active Vancouver: A Year-Round Guide to Outdoor Recreation in the City's Natural Environments - https://rmbooks.com/book/active-vancouver/
The Living Planet Index - https://www.livingplanetindex.org/
Coastal Squeeze - https://thetyee.ca/News/2021/03/17/Coastal-Squeeze/
City of Vancouver Sea Level Rise - https://vancouver.ca/green-vancouver/sea-level-rise.aspx

Bald Eagles at the Chehalis Flats Bald Eagle & Salmon Preserve (Harrison Mills) & at Brackendale (Squamish)

eBird's Status and Trends Abundance Animation - https://science.ebird.org/en/status-and-trends/species/baleag/abundance-map-weekly?week=1
Chehalis Flats Bald Eagle & Salmon Preserve - https://hancockwildlife.org/fraser-valley-bald-eagle-festival/
Brackendale Eagles Provincial Park - https://bcparks.ca/explore/parkpgs/brackendale_eagles/
Hancock Foundation Livestream Eagle Cams - https://hancockwildlife.org/project/live-cams/
Breeding Bird Atlas - https://www.birdatlas.bc.ca/
Breeding Bird Survey - https://www.birdscanada.org/bird-science/breeding-bird-survey
Adopt-a-Nest program - https://hancockwildlife.org/adopt-a-nest/
EagleWatch - https://www.squamishenvironment.ca/programs/eaglewatch/
Mission's Inch Creek Hatchery - https://www.pac.dfo-mpo.gc.ca/sep-pmvs/hatcheries-ecloseries/inch-eng.html
Weaver Creek Spawning Channel - https://www.pac.dfo-mpo.gc.ca/sep-pmvs/hatcheries-ecloseries/spawning-channel-frayeres-artif/weaver-eng.html
Xá:ytem (Hatzic Rock) Canadian National Historic Site - https://www.historicplaces.ca/en/rep-reg/place-lieu.aspx?id=2256

Moose on Green Mountain at Wells Gray Provincial Park

Moose Signs - http://wildwoodtracking.com/mammals/moose/index.html
Wells Gray Protection Society - https://friendsofwellsgraypark.ca/action.html
BC Government's Environmental Protection and Sustainability - https://www2.gov.bc.ca/gov/content/environment/plants-animals-ecosystems/species-ecosystems-

at-risk/implementation/conservation-projects-partnerships/southern-mountain-caribou-rearing-in-the-wild?keyword=southern&keyword=mountain&keyword=caribou
Bowron Lake Canoe Circuit - https://bcparks.ca/explore/parkpgs/bowron_lk/
BC Government Brochures - https://www2.gov.bc.ca/gov/content/environment/plants-animals-ecosystems/species-ecosystems-at-risk/brochures
Washington Department of Fish and Wildlife Fact Sheets - https://wdfw.wa.gov/species-habitats/living/species-facts

Appendices

The Great Acceleration - https://www.anthropocene.info/great-acceleration.html
The Nine Planetary Boundaries - https://www.stockholmresilience.org/research/planetary-boundaries
Extirpation - https://www2.gov.bc.ca/assets/gov/environment/plants-animals-and-ecosystems/species-ecosystems-at-risk/brochures/extinct_extirpated_species.pdf
Northern Spotted Owl - https://www2.gov.bc.ca/gov/content/environment/plants-animals-ecosystems/species-ecosystems-at-risk/implementation/conservation-projects-partnerships/northern-spotted-owl
Southern Mountain Caribou herds - https://www2.gov.bc.ca/assets/gov/environment/plants-animals-and-ecosystems/wildlife-wildlife-habitat/caribou/bc_caribou_herds_population_estimates.pdf
Taylor's Checkerspot Butterfly - https://www2.gov.bc.ca/gov/content/environment/plants-animals-ecosystems/species-ecosystems-at-risk/implementation/conservation-projects-partnerships/taylors-checkerspot

CAN ONE HAVE A FAVOURITE BARNACLE? IF SO, MINE ARE GOOSENECK BARNACLES. I TOOK THIS PICTURE OF THEM IN SEEMINGLY INHOSPITABLE SURF IN PACIFIC RIM NATIONAL PARK. THEY HAVE A LONG FLESHY STEM THAT WOULD BEND OVER WITH EACH CRASHING WAVE, THEN BOUNCE RIGHT BACK UP AGAIN!

GLOSSARY

Adaptation – In relation to species, the process by which a species becomes better fitted to its environment as a result of *natural selection* acting upon genetic variation over several generations.

Alevin – A stage in the salmon life cycle when a newly hatched salmon (or trout) is still obtaining nutrients from its attached yolk sac.

Anthropocene – Derived from the ancient Greek *anthropo* meaning man and *cene* meaning new, the Anthropocene is a proposed geological epoch dating from the time of significant human impact on Earth's geological and ecological systems.

At-risk Natural Area – A designation that indicates many species are threatened or endangered, while at the same time being inadequately protected.

Baleen – A filter-feeding system inside the mouths of certain whales, called baleen whales. These whales open their mouths underwater, taking in huge amounts of water. The whale then pushes the water out through the comb-like baleen, filtering animals such as krill, which are then "licked" off the baleen and swallowed by the whale.

Bat Boxes – Chambered structures that provide space for up to 300 bats to safely roost. They must meet certain criteria, such as size and placement, to be effective.

BC Ecological Reserves – Areas in BC selected to preserve representative and special natural ecosystems, plant and animal species, and phenomena. Scientific research and educational purposes are the principal uses of ecological reserves.

BC Wildlife Act – Written for the protection of BC's wildlife, it proclaims that the ownership of all wildlife in British Columbia is vested in the government, and that people have no right to harm or kill a species, unless in accordance with a permit or licence issued under that act or the Animal Health Act.

Bering Land Bridge – An ancient area of land that now lies beneath the Bering Strait between Alaska and Russia, up to 1000 km wide, and extending south to the Aleutian Islands and north into the Arctic Ocean. The land bridge was part of a large landmass called Beringia, which includes mainly the Yukon, Alaska, and Siberia.

Bioaccumulation – The gradual accumulation of toxic substances such as pesticides, heavy metals, or other chemicals in an organism. Animals that feed higher up the food chain are more likely to have higher concentrations of toxic substances in their bodies.

Biodiversity Hotspots – A global designation given to biologically rich, yet threatened, terrestrial regions.

Biofilm – A complex structure on the surface of the sand that is regularly in contact with the ocean. It consists of colonies of bacteria that secrete a mucilaginous protective coating, which in turn is food for many shorebirds, including the Western Sandpiper, which relies on it for up to half of its energy requirements. Learn more about this important interconnection by reading the *Hakai Magazine* article titled "Flying by the Fat of the Sea."

Bioregional Maps – Maps that set cultural and geographic information into natural boundaries and borders. They are sensitive to non-colonial world views.

Bird Blitz – A bird blitz is an organized gathering of people to identify and count birds in a specific area for a defined amount of time (often 24 hours). These are often open to the public.

Cascading Effect – An inevitable, and sometimes unforeseen, chain of events due to an act, such as resource development, negatively affecting an ecosystem and multiple organisms that live there.

Climate Change – The long-term change in Earth's climate due to human-caused increase in greenhouse gas emissions.

Coastal Squeeze – When sea levels rise in a natural environment, a marsh will migrate inland, staying close to the intertidal zone. But a problem arises in coastal cities when we build infrastructure to protect human development from flooding: the marsh is not able to migrate, and drowns when the water gets too deep (*The Tyee*).

Complexity – Many species vary in size, shape, and colour, and are often found in a wide variety of BC habitats that range from our backyard gardens to coastal bogs, from alpine meadows to the Okanagan pocket desert. The diversity has arisen as species have adapted to have particular affiliations with other specific species, such as butterflies and their associations with certain plants.

Crepuscular Species – Species that are primarily active at dawn or dusk, such as the Common Nighthawk.

DDT – *Dichlorodiphenyltrichloroethane* was a pesticide popular in North America in the 1960s, which had enormous negative impacts on species, especially those at the top of food chains, such as Bald Eagles and Peregrine Falcons. Its use was banned worldwide in 2004 because of its designation as a "Persistent Organic Pollutant" or POP.

Dispersal – When juveniles of a species leave their families or colonies to find a mate and claim a new territory to call their own.

Echinoderms – Marine invertebrate creatures with spiny skin (where "echino" = spiny, and "derm" = skin) and radial symmetry, including sea urchins, sea stars, sand dollars, and sea cucumbers. Many of them can be found in the intertidal zone.

Echolocation – Using individual-produced sound waves to bounce off objects and "read" the reflective echo in order to navigate and find prey. The reflected sounds and their timing provide those that are echolocating with a sound "picture" of what is around them. Used by bats, dolphins, porpoises, and many other marine mammals.

Ecological Resilience – The capacity of an ecosystem or species to respond to a disturbance by resisting damage and recovering quickly.

Ecosystem Services – The services the natural world provides, and that we benefit from. They are broken down into *supporting services,* such as pollination by insects, *provisioning services,* such as wood we use from trees, *regulating services,* such as the flood control offered by wetlands, and *cultural services,* such as the educational, recreational, and aesthetic pleasure that we derive from nature.

Ectotherm – An organism, such as a snake, that regulates its body temperature mainly by exchanging heat with its surroundings, such as the rocks. It is commonly referred to as cold-blooded.

Extirpation – The local extinction of a species that ceases to exist in a specific geographic area, although it continues to exist elsewhere.

Food Webs – Every living organism in an ecosystem is part of multiple predator-prey-plant food chains, and each food chain is one possible path that allows energy and nutrients to move through the ecosystem. All of the interconnected and overlapping food chains in an ecosystem make up a food *web*.

Forage Fish – Fish, such as Pacific Herring, that are preyed on by many larger predators, including larger fish, seabirds, and marine mammals. Forage fish typically feed near the base of food chains, on plankton.

Genetic Analysis – The process of determining and studying the visible (phenotype) or invisible (genotype) inheritable characteristics of an organism. Decoding the DNA of species helps us understand genetic features, relationships, and interconnections. It can also be used to determine evolutionary relationships between species.

Genetic Bottleneck – When all of the members of a certain species have similar genetics and when there are no populations from outside areas to add diversity to the gene pool, this is called a bottleneck. Genetic diversity is considered a requirement for a population to be able to adapt to environmental disturbances.

Genotype – An organism's set of heritable genes (DNA), passed down from parents to offspring. The genes take part in determining the characteristics that are observable in an organism (their "phenotype"), such as hair colour or size.

Grassland Ecosystem – Dominated by grasses, grassland ecosystems in BC generally occur in areas where the climate is hot and dry in summer and cool to cold and dry in winter. The soil is often made up of fine sediments in valley or plateau landscapes. The organisms that live in grasslands have adapted to the dry climatic conditions.

Great Acceleration –A term for the dramatic surge in growth rates across a large range of measures of human activity. It is suggested this acceleration began in the mid-1950s and continues to this day. It is a proposed *age* within the also-proposed Anthropocene *epoch*.

Greenhouse Gases (GHG) – Gases that absorb infrared heat, trapping it in the atmosphere. They can be synthetic, naturally occurring, or increased by human activities, such as burning or releasing fossil fuels. Common GHGs include carbon dioxide, methane, and nitrous oxide. Water vapour is also a GHG.

Hibernaculum – The winter dens of species such as snakes, bats, or bears. The plural is hibernacula.

Hibernation – A cyclic, metabolic state whereby heart rates, respiration, body temperature, and even brainwaves are all reduced for a period of time. It is an adaptation used mainly in winter to reduce energy usage in order to survive.

Holocene Epoch – The geological epoch that began approximately 11,650 years ago (after the last glacial period ended) and continues to this day (or to the beginning of the proposed *Anthropocene epoch*). The Holocene epoch has had a relatively stable climate, allowing for the rise of humans from the period of first agriculture to today.

Humane Hazing – Also known as aversion conditioning, this is the process of creating a negative association between a species, such as a Coyote, and humans, without being cruel to the animal. The idea is to provide people with a reliable and non-lethal "set of tools" to create a healthy boundary between wild animals and themselves.

Hummingbird Banding Stations – Volunteer or employee-staffed stations that place tiny, non-irritating, metal bands on hummingbirds' legs in order to collect data on species health and populations. Banding stations exist for other birds as well.

Hummingbird-friendly Gardens – People can have a positive influence on species by planting gardens that offer improved habitat for certain species. Species-friendly gardens can also be planted for insects, bats, and other bird species.

Icarus Satellite Tags – These are miniature transmitters used to track animals as they move around the globe. The data collected enable scientists to monitor animal migration and to gain more knowledge about the habits and movements of species on our planet.

Individuality – A reference to species individuals that visit an area repeatedly and are the same individual as the last time, as opposed to another organism of the same species. We see people as unique and individual, with particular characteristics distinguishing one from another. This concept can be extended to non-human species that visit our backyards as well.

Intersectional Species – Species that inhabit a niche where urban and wilderness meet. Coyotes, Racoons, Black-tailed Deer, and Black Bears living or frequenting populated areas are all intersectional species.

Invasive Species – Non-native organisms, introduced either actively or passively by humans, that negatively alter their new environment, causing ecological, environmental, and/or economic damage.

Journal Recording – There is great learning in using a journal to record natural observations over the course of a year. See "Phenology," below or in the chapter on winter birds.

Kettling – A term used to describe the behaviour of Turkey Vultures or raptors who use warm currents of air (known as "thermals") to rise high into the air together.

Key Biodiversity Areas – Sites that have been designated as among the most diverse places on Earth for nature; they are sites of global importance to the planet's overall health and the persistence of biodiversity. The Fraser River delta has this designation.

Keystone Species – Species whose role within an ecosystem has a disproportionate effect on other organisms within that system.

Listed Species – These are species that are being monitored provincially, federally, or globally due to their concerningly low population status. In BC they are grouped into three levels of conservation concern: yellow, blue, and red.

Living Planet Index – A measure of the state of the world's biological diversity, based on population trends of vertebrate species from terrestrial, freshwater, and marine habitats. There is a Living Planet Canada report as well.

Mafia Protection – It appears Great Blue Herons nest near Bald Eagles because the nesting eagles are territorial, keeping other predatory birds away from the herons. The downside is the eagle pair see heron chicks as food for their own young. Researchers call it Mafia Protection!

Midden – An archeological term for a pile of refuse left in an area by humans who occupied it in the past, usually containing food leftovers (animal bones, shells, or plant material) along with cultural artefacts, pottery shards, and other miscellaneous items.

Moult – When a species, such as a bird, sheds part or all of its outer covering, such as feathers, and then replaces it with new growth. Moult also happens in other species, such as snakes and crabs.

Movi – *Mycoplasma ovipneumoniae* is a pneumonia-causing bacterium, commonly carried by domestic sheep, which can be transferred to California Bighorn Sheep with fatal consequences.

Murmuration – Large flocks of birds like starlings or sandpipers can fly in apparent synchronization with others, where they appear to be connected together as they twist, turn, and change direction at a moment's notice.

Neonicotinoids ("Neonics") – Used in agriculture to reduce pests, they are neuro-active insecticides. Research shows they are harmful to bees and also to insect-eating wildlife.

Nest Cams – Nest cams (sometimes known as web cams) are digital cameras overlooking bird nests, which allow viewers to watch nesting birds from their computers. Several organizations make digital livestreams available to watch Great Blue Herons, Bald Eagles, Red-tailed Hawks, and even Coola and Grinder, the two Grizzly Bears that hibernate on Grouse Mountain.

Nymph – The larval stage in dragonfly development, which takes place in ponds. They can remain in this stage for three or more years before they mature and take flight.

Pair Bonding – Individual males and females of some bird species (and other species too) form strong pair bonds that can last for their lifetime. Bonded pairs often reconfirm their bond with each other in the lead-up to mating season, with behaviours such as matching head and body movements, calling in unison, or even performing elaborate "dances."

Phenology – The study of biological occurrences, such as migration, breeding, or the leafing out and flowering of plants, in relation to climatic conditions year-over-year.

Phenotype – The observable physical or biochemical characteristics of a species, determined by both genetic makeup and external environmental influences.

Planetary Boundaries – A concept that involves nine Earth system processes, each of which contains an environmental boundary. The boundaries are points beyond which humanity should not push these global-scale processes, because remaining within the boundaries is integral to maintaining the stability and resilience of the planet. Scientists agree that three of the nine have already been crossed, including climate change, biodiversity loss, and the nitrogen cycle.

Pleistocene Period – The geological epoch that lasted from about 2.58 million years ago until 11,700 years ago, spanning the world's most recent period of repeated glaciations.

Pocket Desert – A semi-arid shrub-land ecosystem, known broadly and used specifically for this southern region of the Okanagan Valley in BC and central Washington State. It is centred around the city of Osoyoos.

Ponderosa Pine Biogeoclimatic Zone – BC is divided into 14 broad geographic areas, each having similar patterns of energy flow, vegetation, and soils as a result of their broadly homogenous macroclimates. The Ponderosa Pine Zone occurs in the Lytton-Lillooet area, the lower valley bottoms of the lower Thompson, Nicola, and Similkameen River areas, as well as the Okanagan Lake area, where it extends into the USA.

Precautionary Principle – An approach used to set policy when outcomes are weakly understood and taking too much could be the cause of potentially catastrophic or irreversible events. For example, Canada's Department of Fisheries states that they follow the precautionary principle when setting the catch-limit allotted to the commercial herring fishery.

Proper Feeding of Hummingbirds – Four parts water to one part white sugar only. Do not add dye.

Recruitment – The term used by biologists to describe the process by which new individuals are added to a population, whether by birth and maturation or by immigration. When most Moose calves are taken by predators, for example, it's considered a time of low recruitment.

Redd – A female salmon uses her tail to dig multiple nests (called "redds") in river gravel beds to lay her eggs. A redd can be up to half a metre deep and will be defended by the female after her eggs are laid and covered with gravel.

Refugia – Glaciation separated many species, such as Bighorn Sheep, into isolated ice-free areas, known as "refugia." In Yukon and Alaska, one of these ice-free areas was called *Beringia*.

Rodenticides – Usually anticoagulant-based products designed as blood thinners, which cause animals that ingest them, such as rats, to die from excessive (internal) bleeding.

Ruminants – Animals with four-chambered stomachs. During rest periods, they ruminate, which is the process of regurgitating their food, chewing it again, and re-swallowing it. California Bighorn Sheep are ruminants.

Salmon Enhancement Facilities (Fish Hatcheries) – Often federal government facilities where young fish (usually salmon) are produced and reared, especially for later release in natural waters.

Sea Star Wasting Disease – A disease that affects sea stars and several other echinoderms that appears sporadically, causing mass mortality of those affected. The last major event to hit the BC coast happened in 2013. The disease may be associated with raised water temperatures.

Sexual Dimorphism – This refers to the males and females of the same species exhibiting different characteristics beyond the differences in their sexual organs. This occurs in many animals and may include characteristics such as size, weight, colour, and markings. The size difference between a male and female Bald Eagle is an example of sexual dimorphism, as are the bright colours of many male ducks compared to the less colourful females.

Shifting Baselines – Hypothesized by UBC's Daniel Pauly, the concept suggests that when we measure change in a system, we typically measure against previously known reference points (baselines), which themselves may represent significant changes from even earlier states of the system.

Site Fidelity – When individuals of a species are extremely faithful to their location in the world, such as the feeder you have that brings the same bird to your kitchen window every year, they are considered to exhibit high site fidelity.

Solastalgia – A form of emotional distress or melancholy brought on by environmental change, or what has been lost. It can come from lived experience where one negatively perceives environmental changes that have taken place over time.

Soundscapes – The combination of "geophony," non-biological sounds such as wind or water, "biophony," the sounds generated by organisms such as birdsongs, and "anthrophony," sounds generated by humans such as industrial or shipping acoustic noise.

Species Recovery Management Plans – The action plans developed to help stop or reverse decline, and remove any threats to help ensure long-term survival of species or ecosystems at risk.

Subterranean River Habitats – The vast majority of water in a river system moves slowly through a labyrinth of networks *under* the gravel, cobble, and sand of rivers and their watersheds.

Superpod – A pod is a social group of whales, and a superpod results when multiple pods come together. In the Southern Resident Killer Whale population, J, K, and L pods sometimes join together for a period of time. It is a socialization process that facilitates mating.

Sustainable Development Goals (SDGs) – The set of 17 goals that most countries in the world have committed to work toward between 2015 and 2030. They encompass social, economic, and environmental goals that together provide a blueprint for a sustainable future.

Thermals – Columns of rising air that form an updraft. They are created by the uneven heating of Earth's surface when the Sun warms the ground, which in turn warms the air directly above it. Turkey Vultures ride thermals to higher altitudes before making the migratory crossing from southern Vancouver Island to the Olympic Peninsula, Washington State.

Torpor – A significant reduction in body temperature and metabolic rate, used by some species to conserve energy, such as hummingbirds that go into torpor at night.

Trophic Level – A species' trophic level is the number of steps they are from the bottom of the food chain, which is always a plant. "High trophic level species" are species such as marine mammals that eat "high" on the food chain.

Trophy Hunting – The idea of hunting big-game species for their heads (trophies) versus hunting for subsistence (food). In BC, targets have included Grizzly and Black Bears, Grey Wolves, California Bighorn Sheep, and Moose, among others.

Unusual Mortality Event – Since January 2019, elevated Grey Whale strandings have occurred along the west coast of North America from Mexico through Alaska. This event has been declared an unusual mortality event.

White-nose Syndrome – A disease caused by a white fungus that thrives in cool, moist environments, which has been devastating to bats in many parts of North America, sometimes eliminating more than 90 per cent of a population.

Wildlife Attractants – Items we leave outside, on purpose or accidently, that draw wildlife closer to our homes and to us. These can include pet food, garbage, recyclables, compost, fruit trees, berry bushes, vegetable gardens, bird feeders, barbeques, citronella candles, and even antifreeze, oil, and hot tub covers.

Wildlife Management Areas – Lands or waters designated under the BC Wildlife Act to protect regional or internationally significant fish and wildlife species or their habitats.

Windshield Phenomenon – A term given to the observation that over the past century, decade by decade, fewer dead insects exist to accumulate on the windshields of people's cars. It has been attributed to a global decline in insect populations caused by human activity.

CONSIDER THE MULTIPLE COLOURS OF SNOW GEESE. THIS PAIR SHOWS NOT ONLY THEIR SNOW WHITE FEATHERS, BUT ALSO THEIR BLACK WINGTIPS, AND PINK BILL AND LEGS. THEIR RUSTY FACES HOWEVER, ARE STAINS FROM THE IRON-RICH SOIL THEY FEED IN, POSSIBLY FROM WRANGEL ISLAND, WHERE THEY MIGRATE TO IN FAR EASTERN RUSSIA. PHOTO ERIC GROSS.

NOTES

1 See https://www.nature.com/news/base-the-social-cost-of-carbon-on-the-science-1.21312.
2 Archeological data provide alternative hypotheses on Pacific Herring distribution, abundance, and variability.
3 Interview with Scott Wallace, February 24, 2021.
4 CBC interview with Scott Wallace.
5 Pacificherring.org: Timeline.
6 Pacificherring.org.
7 Raincoast Conservation Foundation: Pacific Herring.
8 Wildwhales.org: Grey Whale.
9 Wildwhales.org.
10 CBC: *The First Peoples: Nuu'chah'nulth*.
11 Interview with Jim Darling, July 7, 2020.
12 Integrative science: Two Eyed Seeing.
13 West Coast N.E.S.T.
14 South Coast Conservation Program: Great Blue Heron *fannini* ssp. Factsheet.
15 Interview with Fernando Selles, May 7, 2020.
16 Interview with Ross Vennesland, May 8, 2020.
17 *Management Plan for the Great Blue Heron Fannini Subspecies*, Species at Risk Act Management Plan Series (Ottawa: Environment Canada, 2016).
18 Interview with Camille Coray, April 27, 2020.
19 *Management Plan for the Great Blue Heron Fannini Subspecies*.
20 Interview with Alison Moran, July 16, 2020.
21 Audubon.org: A Hummingbird's Shining Armor.
22 Rocky Point Bird Observatory.
23 Partners in Flight.
24 City of Vancouver: Official city bird – Anna's Hummingbird.

25 R.S. Harmon, "Geological Development of Panama," in R.S. Harmon (ed.), *The Río Chagres, Panama*, 45–62. *Water Science and Technology Library*, vol. 52 (Dordrecht: Springer, 2005), https://doi.org/10.1007/1-4020-3297-8_4.
26 World of Hummingbirds.com.
27 Interview with Sue Elwell, September 13, 2020.
28 *The Nature of Things*: "Think Like an Animal."
29 Allaboutbirds.org .
30 K.B. Pigg and R.A. Stockey, "The Significance of the Princeton Chert Permineralized Flora to the Middle Eocene Upland Biota of the Okanogan Highlands," *Washington Geol.* 24 (1996): 32–36.
31 Interview with Frank Ritcey.
32 Interview with Trevor Goward, June 18, 2020.
33 Interview with Mike Mueller, June 17, 2020.
34 BC Parks.ca.
35 Bear Smart Community Program.
36 BC Parks Foundation: iNaturalist Project.
37 Okanagan Similkameen Conservation Alliance.
38 BC Ministry of Environment: Species Atlas, Okanagan Region.
39 *Wildlife Viewing in the South Okanagan and Similkameen* brochure.
40 BC Ministry of Environment: Species Atlas, Okanagan Region.
41 Cornell Lab: Seven Simple Actions to Help Birds.
42 Bernie Krause: "The Voice of the Natural World."
43 Interview with Janet Willson, March 2, 2021.
44 Rochelle Baker, "Endangered Vancouver Island Marmots Are Making a Comeback," *National Observer*, November 23, 2020.
45 Interview with Malcolm McAdie, January 11, 2021.
46 Interview with Adam Taylor, January 12, 2021.
47 Marmots.org: Animal Profile.
48 Jon Shepard and Crispin Guppy, *Butterflies of British Columbia* (Vancouver: UBC Press, 2001).
49 Interview with Crispin Guppy, January 28, 2021.
50 Interview with Winnie Hwo, January 12, 2021.
51 Interview with Denis Knopp, January 20, 2021.
52 Interview with Sydney Cannings, February 11, 2021.
53 Robert A. Cannings and Kathleen M. Stuart, *The Dragonflies of British Columbia* (Vancouver: BC Provincial Museum, 1977).
54 Biodiversity of BC: Introduction.
55 Beaty Biodiversity Museum exhibit.
56 Email conversation with Dana Eye, September 8, 2020.
57 Interview with Stephanie Winton, January 11, 2021.
58 BC Environment Ministry, Okanagan Region: Western Rattlesnake.
59 Email communication with Dr. Christine Bishop, September 8, 2020.
60 BC Reptiles: Northern Pacific Rattlesnake and Gopher Snake.
61 BC Environment Ministry: Wildlife Habitat Areas.
62 BC Environment Ministry: Okanagan Habitat Atlas.
63 Interview with Jo Hughes, August 20, 2020.

64 "Where Did the Columbian Ground Squirrels Go?" Manning Park Resort blog post, August 27, 2020.
65 Owlcation: The Columbia Ground Squirrel in Manning Park.
66 Interview with Kelly Pearce of Hope Mountain Centre, August 19, 2020.
67 Line Cordes, "How Animals Are Coping with the Global 'Weirding' of the Earth's Seasons," *The Conversation.com*, July 8, 2020.
68 "Call to Restrict Neonicotinoids," Letter to the Editors, *Science Magazine,* June 1, 2018.
69 BEEPS bat house interpretive trail panels.
70 Kelowna Museum exhibits: "Bats: Out of the Darkness."
71 Wildlife Conservation Society: Bats.
72 Interview with Darlene Hartford, October 15, 2020.
73 Brooke Jarvis, "The Insect Apocalypse Is Here. What Does It Mean for the Rest of Life on Earth?" *The New York Times*, November 27, 2018, https://www.nytimes.com/2018/11/27/magazine/insect-apocalypse.html.
74 Dirk Sanders, Enric Frago, Rachel Kehoe, Christophe Patterson, and Kevin J. Gaston, "A Meta-Analysis of Biological Impacts of Artificial Light at Night," *Nature Ecology & Evolution* 5 (2021): 74–81.
75 Urban Coyote Research: Home Ranges of Individuals.
76 Interview with Kristen Walker, February 16, 2021.
77 Stanley Park Ecology Society: Co-Existing with Coyotes – Understanding Urban Coyotes.
78 Interview with Dannie Piezas, October 27, 2021.
79 Urban Coyote Research Project: General Information.
80 Urban Coyote Research Project, Research Paper: "Spatial and Temporal Variation in the Diet of Coyotes in the Chicago Metropolitan Area."
81 Nativemaps.org: An Introduction to Bioregional Mapping.
82 Interview with Dan Straker, February 2, 2021.
83 Interview with Phil Lavoie, March 5, 2021.
84 "Red Sea Urchins Discovered to Be One of the Earth's Oldest Animals," Oregon State University, *Newsroom* blog post, July 13, 2009.
85 Lakshmi Supriya, "Sea Urchins Can Drill Holes in Solid Rock with Just Their Teeth," *New Scientist*, February 21, 2018.
86 Rob Monroe, "How Much CO2 Can the Oceans Take Up?" *Keeling Curve* blog post, July 3, 2013.
87 Bamfield Marine Sciences Centre: BMSC Overview.
88 United Nations Development Programme, Sustainable Development Goals: SDG 14 Life Below Water.
89 Raincoast Conservation Foundation: Salish Sea, Chapter 4 – The Tourist Dollar.
90 Parks Canada Interpretive Sign at East Point.
91 Marine Mammal Commission: Southern Resident Killer Whale.
92 Whale Research.com: Orcas and Salmon.
93 Larry Pynn, "A Whale for the Living: Many Species Continue to Struggle, but Marine Mammals Are Making a Comeback," *The Tyee*, April 2012.
94 Interview with Donna Sandstrom, in Chole Berge, "Humans Are So Loud, Marine Safaris Might Soon

Be on Land," *The Daily Beast,* April 25, 2021.

95 The Whale Trail: East Point, Saturna Island.

96 Interview with Lucy Quayle, August 5, 2020.

97 Karin Larsen, "Coast Guard Throws Technology Behind Protecting Southern Resident Killer Whales," *CBC News,* January 14, 2021.

98 "What's Killing Killer Whales? Orca Report Covering a Decade of Necropsies Identifies Threats," *UBC News,* December 2, 2020.

99 HappyWhale.com.

100 See https://sdgs.un.org/goals/goal4.

101 SIMRES: Marine Noise & SRKW Research.

102 Robert M. Wilson, *Seeking Refuge: Birds and Landscapes of the Pacific Flyway* (Seattle: University of Washington Press, 2010).

103 P. Stamets, *Psilocybin Mushrooms of the World: An Identification Guide* (Berkeley, CA: Ten Speed Press, 1996).

104 Armel Castellan, Environment and Climate Change Canada.

105 *Indicators of Climate Change for British Columbia: 2016 Update*. BC Ministry of Environment.

106 G. Albrecht, "Solastalgia: A New Concept in Human Health and Identity," *PAN: Philosophy, Activism, Nature* 3 (2005): 41–55.

107 All About Birds: Turkey Vulture Range Map.

108 BirdWeb.org: Turkey Vulture.

109 Larry Pynn, "How Do You Feel about Turkey Vultures?" *The Tyee,* May 21, 2020.

110 D.E. Alexander, "Bioaccumulation, Bioconcentration, Biomagnification," in *Environmental Geology: Encyclopedia of Earth Science* (Dordrecht: Springer, 1999), https://doi.org/10.1007/1-4020-4494-1_31.

111 See orcarelief.org: P.S. Ross et al., "High PCB Concentrations in Free Ranging Pacific Killer Whales," *Marine Pollution Bulletin* 40, no. 6 (2000): 504–15.

112 Audubon.org: Survival by Degrees.

113 Interview with Ann Nightingale, September 25, 2021.

114 Bob Turner's Chum spawn YouTube video.

115 Larry Pynn, "Salmon Smolts: Here Today, Guano Tomorrow," *Hakai Magazine,* February 26, 2021.

116 Jim Robbins, "A New Way of Understanding What Makes a River Healthy," *Yale e360*, January 4, 2017.

117 GoldstreamPark.com: Salmon Run.

118 Sannich.ca: Indigenous Relations.

119 See Pacific Salmon Foundation Community Salmon Program Fact Sheet.

120 Raincoast Conservation Foundation: "Increasing salmon hatcheries could do more harm than good for Chinook and Southern Resident killer whales."

121 Animal Diversity Web: *Ovis canadensis* – Bighorn sheep.

122 Gale Monson and Lowell Sumner (eds.), *The Desert Bighorn: Its Life History, Ecology & Management* (Tucson: University of Arizona Press, 1980).

123 Bighorn Institute: Wild Sheep of North America.

124 *Hinterland Who's Who*: Bighorn Sheep.

125 Interview with Edyta Marcisz, February 12, 2021.

126 Protect the Grasslands of BC Interpretation Panel – Lac du Bois.

127 Interview and walk with Frank Ritcey, October 6, 2020.

128 Raincoast Conservation Foundation: Guidelines for Ethical Photography.

129 Chris Darimont et al., "Large Carnivore Hunting and the Social Licence to Hunt," *Conservation Biology* 35, no. 4 (2020): 1111–19, https://doi.org/10.1111/cobi.13657.

130 Key Biodiversity Areas: About KBAs.

131 All About Birds: Barrow's Goldeneye.

132 Globally Threatened Species Bird Forum: Barrow's Goldeneye.

133 David P.L. Toews and Darren E. Irwin, "Cryptic speciation in a Holarctic passerine revealed by genetic and bioacoustic analyses," *Molecular Ecology* 17, no. 11 (2008): 2691–2705.

134 R. Butler and G. Clulow, "History of the Establishment of a Large Northwestern Crow Roost in Burnaby, British Columbia," *British Columbia Birds* 30 (2020): 7–9.

135 Cornell Lab BirdCast: About.

136 Mark L. Hineline, *Ground Truth: A Guide to Tracking Climate Change at Home* (Chicago: University of Chicago Press, 2018).

137 Michelle Gamage, "Coastal Squeeze," *The Tyee*, March 17, 2021.

138 Interview with David Hancock, January 25, 2021.

139 Fraser Valley Bald Eagle Festival: Announcing the Chehalis Flats Bald Eagle and Salmon Preserve.

140 Interview with David Hancock, January 25, 2021.

141 Interview with Kristen Walters, Centre for Wildlife Ecology research group at Simon Fraser University.

142 Interview with Judith Knapp, February 8, 2021.

143 BC Factsheet: Moose in BC.

144 Art Rodgers, *Moose* (Stillwater, MN: Voyageur Press, 2001), 24–27.

145 Ian Hatter et al., "Moose Population Dynamics during 20 Years of Declining Harvest in British Columbia," *Alces: A Journal Devoted to the Biology and Management of Moose* 54 (2018): 101–19, https://alcesjournal.org/index.php/alces/article/view/240.

146 R.A. Gorley, *A Strategy To Help Restore Moose Populations In British Columbia: Recommendations Prepared for the Ministry of Forests, Lands and Natural Resource Operations Fish and Wildlife Branch*, 2016.

147 Environmental Reporting BC: *Long-term Change in Air Temperature in B.C. (1900–2013)*.

148 Harriet V. Kuhnlein and Murray M. Humphries, "Traditional Animal Foods of Indigenous Peoples of North America: Moose," Centre for Indigenous Peoples' Nutrition and Environment, McGill University.

149 BC Government: Provincial Moose Research Project.

150 BC Government: Northern Spotted Owl Recovery & Breeding Program.

151 Stan Boutin and Evelyn Merrill, "A Review of Population-Based Management of Southern Mountain Caribou in BC," Report Prepared for Columbia Mountains Institute, March 2016.

152 BC Government: Southern Mountain Caribou.

153 *Wikipedia* was used as a primary resource in defining many of the glossary terms in this list.

THIS GREAT PORTRAIT IMAGE OF A SEA LION PAIR CLEARLY SHOWS THEIR EXTERNAL EAR FLAPS, A CHARACTERISTIC OF THEIR GROUP CALLED PINNIPEDS. PHOTO MICHAEL SCHMIDT PHOTOGRAPHY

INDEX

Note: Page numbers in *italics* indicate figures

About the Author

Roy Jantzen is an environmental educator, naturalist, and author of *Active Vancouver: A Year-Round Guide to Outdoor Recreation in the City's Natural Environments* (2015). He holds a MA in Environmental Education and Communication from Royal Roads University and has had a career as an educator at Capilano University, where he teaches courses in natural history, ecotourism, and climate change. For over three decades, Roy has helped educate the public about the importance of biodiversity and our human place within it. Though he has a passion for local ecosystems and the species that inhabit them, he also has a strong desire to relate the environment to our lives and larger issues such as the Sustainable Development Goals, planetary boundaries, and our personal ecological footprints. Roy is passionate about helping his students become the next generation to answer the call of creating a sustainable planet, and asks, "Shouldn't all education be environmental education?"